Cannabinoids and the Brain

Cannabinoids and the Brain

Linda A. Parker

The MIT Press
Cambridge, Massachusetts
London, England

First MIT Press paperback, edition, 2018

Set in Sabon LT Std by Toppan Best-set Premedia Limited. Printed and bound in the United States of America.

Library of Congress Cataloging-in-Publication Data

Names: Parker, Linda (Linda A.), author.
Title: Cannabinoids and the brain / Linda Parker.
Description: Cambridge, MA : The MIT Press, [2017] | Includes bibliographical references and index.
Identifiers: LCCN 2016030736 | ISBN 9780262035798 (hardcover : alk. paper)
Subjects: LCSH: Nervous system--Degeneration--Treatment. | Cannabis--Therapeutic use. | Marijuana--Therapeutic use.
Classification: LCC RM666.C266 P37 2017 | DDC 615.7/827--dc23 LC record available at https://lccn.loc.gov/2016030736

10 9 8 7 6 5 4 3 2

for Ernie

Contents

Preface

My keen interest in the endocannabinoid system began with an airmail envelope from Israel that arrived in my faculty mailbox in the year 2001. In late 1999, Cheryl Limebeer and I published a paper showing that THC reduced nausea in a rat model that we had had a role in developing. Shortly after the paper was published, I received a letter from Professor Raphael Mechoulam suggesting that we test non-psychotropic cannabidiol in that model. I was thrilled! The letter began a very exciting and fruitful collaboration that brought me into the exciting world of endocannabinoids. In 2002, at Professor Mechoulam's kind invitation, I attended my first meeting of the International Cannabinoid Research Society, in Monterey, California, at which I met Professor Mechoulam and several other outstanding researchers who would later become collaborators.

In addition to Professor Mechoulam, our group has had the good fortune of collaborating with such leaders in their field as Roger Pertwee (University of Aberdeen), Keith Sharkey (University of Calgary), Daniele Piomelli (University of California at Irvine), Aron Lichtman (Virginia Commonwealth University), Benjamin Cravatt (Scripps Research Institute), Loren Parsons (Scripps Research Institute), Alexandros Makriyannis (Northeastern University), Paul Fletcher (Center for Addiction and Mental Health), Bryan Kolb (University of Lethbridge), Klaus-Peter Ossenkopp (University of Western Ontario), John Salamone (University of Connecticut), Ganesh Thames (Northeastern University), and Matias Lopez (University of Oviedo), and with my wonderful colleagues at the University of Guelph, Elena Choleris, Francesco Leri, Boyer Winters, Bettina Kalisch, and Craig Bailey. Over the past ten years these collaborations have provided outstanding training opportunities for members of my laboratory, including Cheryl Limebeer, Erin Rock, Martin Sticht, Kiri Wills, Lesley O'Brien, Shadna Rana, Katherine Tuerke, and Gavin Petrie.

It is these hard-working, enthusiastic, highly skilled people who keep the process of discovery going in the laboratory. Nothing would happen without them.

This book is an expansion and update of a review paper, which Raphael Mechoulam kindly invited me to co-author, that appeared in the *Annual Review of Psychology* (Mechoulam and Parker 2013). Our collaborations over the past 15 years have always taken my research to new vistas. Astute and careful editing by Paul Bethge at the MIT Press has made the book a much better read. I thank him for his contributions to the clarity of my prose.

I draw the reader's attention to four outstanding edited volumes on cannabinoids and endocannabinoids that have been published in the past few years: *Handbook of Cannabis*, edited by R. G. Pertwee (Oxford University Press, 2014), *Handbook of Experimental Pharmacology*, volume 231: *Endocannabinoids*, edited by R. G. Pertwee (Springer, 2015), *International Review of Neurobiology*, volume 125: *Endocannabinoids*, edited by L. Parsons and M. Hill (Elsevier, 2015), and *Cannabinoids*, ed. V. Di Marzo (Wiley-Blackwell, 2014). For more information, see the Information for Health Care Professionals on the Health Canada website at http://www.hc-sc.gc.ca/dhp-mps/marihuana/med/infoprof-eng.php.

The National Institutes of Health recently prepared a videocast of a two-day conference (Marijuana and Cannabinoids: A Neuroscience Research Summit) held in March of 2016 and featuring presentations by some of the foremost NIH-funded researchers. I encourage readers of this book to find a free evening, open the link (http://videocast.nih.gov), and hear about the current state of the research in this exciting field. I also recommend *The Scientist*, a video documentary (produced by Fundación CANNA and available on YouTube) that provides an glimpse into the scientific and personal life of Raphael Mechoulam, "the father of the endocannabinoid system," a humble, brilliant, generous, and truly delightful human being. I have had the great fortune of being able to call Professor Mechoulam my friend.

1

An Introduction to Cannabis

The cannabis plant has been used for recreational and medicinal purposes for more than 4,000 years, but only recently has scientific investigation of its effects begun to be fruitful. In 1964, at the Hebrew University of Jerusalem, the chemists Yehiel Gaoni and Raphael Mechoulam discovered the principal psychoactive ingredient in cannabis, Δ^9-tetrahydrocannabinol (often referred to as THC) (Gaoni and Mechoulam 1964a). This discovery finally provided the scientific community with the tools needed to investigate the properties of this plant. Although it was clear that the psychoactive effects of THC were produced by its action on the brain, exactly how that happened was not discovered until 20 years later, when Allyn Howlett's group at the St. Louis University Medical School discovered the cannabinoid-1 (CB_1) receptors (Devane et al. 1988; Howlett et al. 1986). The brain distribution of these newly discovered receptors was found to be consistent with the known pharmacological effects of THC. This discovery raised a question: Why would the brain have receptors for a material derived from a plant? Obviously, the brain must produce chemicals similar to THC. Shortly thereafter, Mechoulam's group discovered anandamide (AEA) (Devane et al. 1992) and 2-arachidonyl glycerol (2-AG) (Mechoulam et al. 1995; Sugiura et al. 1995), which are called *endocannabinoids* because they act on the same receptors as THC. And a second cannabinoid receptor was discovered: CB_2, found primarily in the immune system (Munro, Thomas, and Abu-Shaar 1993). The discovery of the site of action of THC has uncovered a new brain chemical system that is intimately involved in all aspects of brain functioning.

Since the aforementioned discoveries, a large number of investigators have provided considerable scientific knowledge about the metabolism, biochemistry, and pharmacology of THC, mostly in animal models. In a recent comprehensive review, Pacher and Kunos (2013) state that "modulating endocannabinoid activity may have therapeutic potential in almost all diseases affecting humans, including obesity/metabolic syndrome, diabetes and diabetic complications, neurodegenerative, inflammatory, cardiovascular, liver, gastrointestinal, skin diseases, pain, psychiatric disorders, cachexia, cancer, chemotherapy induced nausea and vomiting, among others." This bold statement is based on considerable pre-clinical scientific evidence in each of these fields. This newly discovered system is of great importance and can explain many of the medicinal claims that have been made for medical marijuana. However, because of legal constraints, very little clinical work has been done in humans (Mechoulam et al. 2014).

In modern medicine there are few approved conditions for the use of cannabis and cannabinoid-derived medicines. This is due to the rather limited high-quality human clinical trials with cannabinoids. To be approved for use in a patient population, a drug must undergo clinical trials. Clinical trials have three phases to assess a new medicine for safety and effectiveness. Phase I tests for safety. A small number of people, are given a small dose of the drug under careful supervision. This is not done to test whether the drug works for the condition under review; it is done to test for side effects. In phase II the drug is given to people who have a medical condition to determine whether or not it helps them. Phase III trials are conducted only for medicines that have already passed the first two phases. In phase III, the drug is compared against existing treatments and/or placebo, using randomized assignment to treatment groups (randomized control trials, abbreviated RCT) and double-blind procedures (neither participant nor the researcher knows to which group a participant was assigned). These clinical trials often last a year or longer, involving several thousand patients and often millions of dollars.

According to a recent systematic review and meta-analysis (Whiting et al. 2015) of evidence (a total of 79 RCTs) of cannabinoids as medicine for nausea/vomiting, appetite stimulation, chronic pain, spasticity in multiple sclerosis (MS), depression, anxiety disorder, sleep disorder, psychosis, glaucoma, or Tourette syndrome, most studies suggested that cannabinoids were associated with improvements in symptoms. However, when the quality of each study was assessed using the Cochrane risk-of-bias tool, which grades the rigor of the design of a clinical trial, there was moderate-quality evidence to support the use of cannabinoids in the treatment of chronic neuropathic or cancer pain and spasticity due to MS. There was low-quality evidence suggesting that cannabinoids were associated with improvements in nausea and vomiting due to chemotherapy, weight gain in HIV infection, sleep disorders, and Tourette syndrome, and very low-quality evidence for an improvement in anxiety. A number of methodological weaknesses were identified in these studies, including failure to appropriately handle participants who withdrew from the trial, selective outcome reporting, inadequate description of methods, inadequate randomization, and inadequate blinding. An additional limitation was the very small sample size of most of the RCTs, which means that they lacked the power to detect any difference in the treatments. Multiple different cannabinoids and routes of administration also were evaluated in various studies. In view of these limitations, Whiting et al. urged that large, robust RCTs be conducted to confirm the potential medicinal effects of cannabinoids. However, such trials are expensive and are difficult to perform with substances that are highly regulated. Therefore, most of the evidence of the potential medicinal promise of cannabinoids is based on experimental pre-clinical studies with animals, in which greater scientific rigor can be applied to the design of a study. The numbers of such studies have increased exponentially since the discovery of the endocannabinoid system and the mechanism of action of cannabinoids in the 1990s. Results clearly demonstrate the promise of cannabinoids for treatment of several disorders of the nervous system, including anxiety, post-traumatic stress disorder (PTSD), chronic pain, spasticity in MS, epilepsy, obesity, nausea, and vomiting.

A Brief History of Cannabis Use

Early in agricultural history, more than 10,000 years ago, cannabis was grown along the rich banks of great rivers, including the Hwang-Ho in China, the Indus in India, the Tigris and the Euphrates in Mesopotamia, and the Nile in Egypt, for the purpose of producing hemp fibers. The oldest known written record of cannabis use comes from the Chinese emperor Shen Nung, who in 2,727 BC investigated botanical compounds for the potential medicinal properties, using himself as a test subject (Abel 1980). As the "father of Chinese Medicine," Shen Nung is said to have written the original *Pen Ts'ao*, listing hundreds of drugs derived from vegetable, animal, and mineral sources; however, no original text exists. The oldest *Pen Ts'ao* was compiled in the first century AD by an unknown author who claimed to have incorporated the original compendium into his book, which contains a reference to *ma*, the Chinese word for cannabis. In the second century AD, the Chinese surgeon Hua T'o used a mixture of cannabis resin and wine as an anesthetic to perform painless surgery. The Chinese were also the first people to report the use of recreational cannabis (Abel 1980).

India was the first culture to use cannabis for religious and social purposes. In the four holy books known as the Vedas, the god Siva became angry after a family squabble and takes refuge in the cool shade of a cannabis plant. He eats some of the plant, and it becomes his favorite food. Siva became known as the Lord of Bhang (Abel 1980). Bhang is not merely the plant, but also a liquid refreshment made with its leaves. Ganja and charas, too, were made from cannabis in India. Ganja, made from the flowers, is more potent than bhang. Charas is made from the resin of the flowers and is much more potent than ganja; it is similar to hashish. Bhang is used in a manner similar to alcohol in the West—often in conjunction with weddings and special festivities. According to Indian folk songs dating back to the twelfth century, bhang was also consumed by soldiers before they went into battle. At the turn of the twentieth century, the Indian Hemp Drug Commission, summoned to investigate the use of cannabis in India, concluded that the plant was an integral part of the culture and religion of the country. Indeed, the commission concluded that "besides a cure for fever, bhang has many medicinal virtues. It cures dysentery and sunstroke, clears phlegm, quickens digestion, sharpens appetite, makes the tongue of the lisper plain, freshens the intellect, and gives alertness to the body and gaiety to the mind."

Cannabis was introduced into Western medicine in the 1839 by William Brooke O'Shaughnessy (1839), a surgeon who learned of its medicinal benefits while working for the British East Indies Company. It was promoted for analgesic, sedative, anti-inflammatory, anti-spasmodic, and anti-convulsant properties (properties for which pre-clinical animal models provide considerable support). In fact, cannabis was said to be the treatment of choice for Queen Victoria's menstrual pain (Abrams and Guzman 2015). However, in the 1900s specific medicines were developed for each of the above-mentioned indications, and medicinal use of cannabis declined in Western countries.

In the middle of the nineteenth century, Charles Baudelaire, Théophile Gautier, and Jacques-Joseph Moreau—all members of the Parisian Club des Hachichins—wrote about the psychological effects of cannabis (Mechoulam and Hanus 2000). In 1845, Moreau, a psychiatrist, published a book, titled *Hashish and Mental Illness*, in which he described psychological phenomena he had observed in experimental participants, including happiness, excitement, dissociation of ideas, distortion of time and space, sensory enhancement, delusions, emotional fluctuations, illusions, and hallucinations. It is presumed that Moreau's participants consumed large amounts of hashish orally (Mechoulam and Parker 2013).

In the United States, the history of the use and the prohibition of cannabis have been clearly described many times (Brecher 1972; Iversen 2008). Early in the twentieth century, marijuana was used for its psychoactive properties mostly by Mexicans and by African American jazz musicians in a few large cities. In fact, the wave of immigrants who entered the southern US from Mexico in that period first brought the drug under suspicion by authorities and eventually led to its prohibition. Its recreational use first became widespread in New Orleans, then gradually spread to other cities. In a campaign to outlaw the drug waged by the head of the Federal Bureau of Narcotics, Harry Anslinger, it was alleged that marijuana use led to violent crimes. In the late 1930s, newspaper stories in large cities were filled with stories about "killer weed." In 1937 the US Congress passed the Marijuana Tax Act, which effectively banned the use of the drug in medicine and outlawed it as a dangerous narcotic, effectively terminating research on it for about 25 years.

International Control of Cannabis

Late in the nineteenth century, the British Parliament required the colonial Government of India to establish a Royal Commission on Opium and an Indian Hemp Drugs Commission. The latter concluded that moderate use of cannabis had no appreciable physical effects on the body, no harmful effects on the brain, and no adverse effects on morality (Abel 1980; Mills 2003). It recommended a system of taxation, control, and restriction, but not prohibition (Mead 2014; Mills 2003). Despite the reports, use of cannabis as a medicine faded as improved technology facilitated the study of pharmacology and organic chemistry, the isolation of active ingredients from natural products, and the synthesis of new molecules (Anderson 2005). Morphine was identified and isolated in 1805 by Friedrich Sertürner, and the subsequent development of the hypodermic needle provided a means of administering it for rapid pain relief. Later in the nineteenth century, aspirin and other medications (among them barbiturates) were developed. Complex, unrefined herbal medicines became of less interest to mainstream medicine. When efforts to control the use of cannabis arose in Europe and North America in the middle of the twentieth century, the medical community did not defend its use as a medicine (Mead 2014).

Early in the twentieth century there was a growing interest in international cooperation in the control of opium. During the Second Opium Conference, held in Geneva in 1925, cannabis and the resin used to produce hashish also were discussed (Mead 2014). In 1954, the World Health Organization reported that cannabis had few medical benefits. In 1961, an international treaty called the Single Convention on Narcotic Drugs established limits on the production, manufacture, and extraction of raw materials from what were determined to be narcotic drugs. Psychoactive substances (drugs) were placed in four schedules, the most restrictive of which contained drugs viewed to be particularly dangerous for abuse liability and to have little therapeutic value. The cannabis plant, its resin, its extracts, and its tinctures were placed in this most restrictive schedule. Subsequently, at the 1971 UN Convention on Psychotropic Substances, THC and its isomers were placed in Schedule I, which prohibited all use except for scientific purposes and very limited medical purposes by duly authorized persons specifically approved by a governmental body. Other cannabinoids (among them cannabidiol) were not controlled under the 1971 UN Psychotropic Convention. Those non-psychoactive cannabinoids are still not controlled by many countries (e.g., Great Britain), but the United States and Canada have chosen to include them in the same restrictive schedule as THC. (For an excellent discussion of how various nations around the world have interpreted and implemented the UN treaty and have devised regulatory tools to distinguish between medical and non-medical use, see Mead 2014.)

Constituents of Cannabis

Although more than 100 plant cannabinoids (phytocannabinoids) have been identified in the cannabis plant, there is essentially only one compound that causes the typical "marijuana" effects: THC. However, other compounds (including cannabidiol, abbreviated CBD) can modify the effects of THC. The sought-after euphoria produced by smoking marijuana is entirely attributable to THC and not at all attributable to the other cannabinoids found in the plant (Mechoulam et al. 2014).

THC is not evenly distributed throughout the plant. It is absent from the roots and the seeds and is found only at very low concentrations in the stems. The lower leaves contain less THC than the upper leaves (typically 2–3 percent). However, unpollinated, all-female floral material—called *sinsemilla*, meaning "without seeds" in Spanish—is a primary source of THC (with concentrations as high as 25 percent) and other cannabinoids. However, the concentration of cannabinoids within individual plants and tissues is due mainly to the presence or absence of glandular or capitate trichomes—small structures in which cannabinoids are synthesized and sequestered (Potter 2014).

Varieties of cannabis plants differ in their cannabinoid contents. The concentration of THC in industrial hemp is less than 0.3 percent. In hashish in the 1960s it was about 5 percent, whereas in marijuana it was about 2–3 percent. In the twenty-first century, strains of marijuana have been developed that contain up to 25 percent THC.

Because all its types are able to interbreed, many academic botanists argue that cannabis is one polymorphic species (Piomelli and Russo 2016; Russo 2014), and that the biochemical cannabinoid and terpenoid profiles of different strains cannabis account for their differing effects. On the other hand, there is recent genetic evidence indicating that hemp is genetically more similar to *Cannabis indica* than to *Cannabis sativa* (Sawler et al. 2015).

The cannabinoids identified from cannabis are of eleven types (ElSohly and Gul 2014; Mechoulam 2005): Δ^9-THC, Δ^8-THC, cannabidiol (CBD), cannabigerol (CBG), cannabichromene (CBC), cannabinodiol (CBND), cannabielsoin (CBE), cannabicyclol (CBL), cannabinol (CBN), cannabitriol (CBT), and miscellaneous cannabinoids. In addition to cannabinoids, the cannabis plant contains hundreds of non-cannabinoid constituents, including terpenoids, which give the plant its characteristic smell (and which are also found in hops, a close relative to cannabis). *In vitro* and *in vivo*, some terpenoids have been found to have anti-inflammatory, anti-bacterial, and anti-anxiety effects (Piomelli and Russo 2016); however, no clinical trials support those claims. Various combinations of cannabinoid constituents and their interaction with terpenoids could provide diverse therapeutic results, accounting for why people claim different symptomatic relief from using different strains. However, all the evidence is anecdotal. Human clinical trials will be needed to harvest the "treasure trove" (Mechoulam 2005) of compounds found in the cannabis plant.

THC-type cannabinoids

Of the various components, THC types have received the most scientific investigation. The effects of Δ^9-THC (commonly called simply THC) will be highlighted throughout this book, because it is the primary psychoactive cannabinoid found in cannabis. Very small quantities of Δ^8-THC also are found in cannabis. Mechoulam et al. (2014) isolated numerous cannabinoids from Lebanese hashish that had been confiscated by Israeli police in Jerusalem in the 1960s and elucidated their structures. Those compounds were then tested for cannabinoid activity in monkeys, and only Δ^9-THC was found to produce the typical psychoactive effects of marijuana. Once Δ^9-THC had been isolated, Roger Pertwee (1972) developed the *ring test*, a quantitative *in vivo* assay for psychotropic cannabinoids based on how long a mouse placed across a horizontal ring remains immobile or cataleptic. Later, Billy Martin et al. (1991) used the ring test and three other bioassays in what came to be known as the *mouse tetrad assay*; it included, catalepsy, hypokinesia (inactivity), hypothermia (reduced body temperature), and anti-nociception (pain relief). The mouse tetrad assay is used to screen for psychotropic cannabinoids, all of which, in contrast to many other classes of drugs, generally show similar potency in all four of these behavioral measures (Mechoulam et al. 2014).

After the identification of THC, several investigations on humans showed that, when THC was taken orally or intravenously or when it was inhaled in smoke, it showed substantial potency similar to that reported by recreational cannabis users. At present, two analogues of THC that have been approved by the US Food and Drug Administration are available in the United States in capsules that may be prescribed for chemotherapy-induced nausea and vomiting: nabilone (Cesamet, Valeant Pharmaceuticals North America) and dronabinol (Marinol; Solvay Pharmaceuticals). More recently, the cannabis-based medicine nabiximols (Sativex, GW Pharmaceuticals), which contains approximately equal amounts of THC and CBD and is administered as a sub-lingual spray, was first licensed as a medicine in Canada for relief of pain in adult patients suffering from multiple sclerosis (MS) or cancer and subsequently to reduce spasticity in MS patients (Mechoulam et al. 2014).

Other Δ^9-THC-type compounds in the plant are not psychoactive but may have medical uses. The carboxylic acidic precursor of THC is Δ^9-THC-acid (THCA) (Gaoni and Mechoulam 1964b). In the fresh plant, THCA is decarboxylated to THC by heating or burning. Interestingly, no psychomimetic activity was observed when THCA was administered to rhesus monkeys at doses up to 5 milligrams per kilogram (intravenously, abbreviated iv), to mice at doses up to 20 mg/kg (intraperitoneally, abbreviated ip), and to dogs at doses up to 7 mg/kg (Grunfeld and Edery 1969). THCA has recently been reported to inhibit the pro-inflammatory cytokine tumor necrosis factor α (TNF-α) *in vitro* (Verhoeckx et al. 2006).

THCA also serves as an anti-emetic and anti-nausea agent in animal models, which suggests that it may have therapeutic potential (Rock, Kopstick, Limebeer, and Parker 2013).

Tetrahydrocannabivarin (THCV), identified in the 1970s (Gill 1971; Merkus 1971), was initially described as sharing with Δ^9-THC the ability to produce catalepsy in the mouse ring mobility test and as able to produce mild Δ^9-THC-like effects in humans (Hollister 1974). These effects have been shown to be limited to very high doses, with low doses acting as an antagonist at the CB_1 receptor (Pertwee et al. 2007). Indeed, at low doses THCV suppresses food intake and reduces body weight, as does the CB1 receptor antagonist/inverse agonist rimonabant (a drug previously approved for weight control in humans). However, unlike rimonabant, THCV produces neither nausea (Rock, Sticht, Duncan, Stott, and Parker 2013) nor anxiety (O'Brien, Limebeer, Rock, Bottegoni, Piomelli, and Parker 2013), two side effects produced by inverse agonism of the CB_1 receptor that limit the usefulness of rimonabant as a drug. Therefore, THCV may be a useful drug for weight loss.

Cannabidiol-type cannabinoids

The primary non-psychoactive cannabinoid of cannabis (particularly in hemp) is cannabidiol (CBD), first isolated from Mexican marijuana by Roger Adams (1941) and from Indian Charas by Alexander Todd (1946). Mechoulam and Shvo (1963) later isolated CBD from Lebanese hashish and established its structure. Unlike THC, CBD lacks psychotropic activity; however, research from pre-clinical animal models suggests that it has therapeutic potential for management of inflammation, anxiety, emesis, nausea, inflammatory pain, and epilepsy. In addition, as will be discussed in later chapters, CBD may reduce the psychoactive and memory-impairing effects of THC. CBD also produces neuroprotective and anti-oxidant effects (Mechoulam, Parker, and Gallily 2002; Pertwee 2008).

In the past twenty years, the THC content of street cannabis has risen dramatically, whereas the CBD content has remained the same or decreased to negligible levels. For instance, in the United States the THC content of street cannabis rose from 4 percent in 1995 to 12 percent in 2014 (ElSohly et al. 2016). Thus, the cannabis sold today differs considerably from the cannabis that was available years ago in its effects on mental health and cognitive function. In the United States, the cannabis that the National Institute of Drug Abuse supplies to researchers for experiments generally has contained less than 4 percent

THC (although various strains are now becoming available to researchers), and thus findings from their experiments may have limited implications for users of present-day street cannabis. As we will see throughout this book, the ratio of THC to CBD in cannabis can dramatically affect the drug's effect on cognitive functions, its abuse potential, its psychotic side effects, and some of its medicinal properties.

Together with Δ^9-THC, CBD is a major constituent of nabiximols (Sativex 2.7 mg THC: 2.5 mg CBD), a medicine developed by GW Pharmaceuticals that is approved in Canada for treatment of cancer pain and for relief of neuropathic pain and spasticity in multiple sclerosis. GW Pharmaceuticals has also developed a CBD sub-lingual spray (called Epidiolex); it is undergoing phase III clinical trials for childhood epilepsy.

Cannabidiolic acid (CBDA) is the non-psychoactive precursor of CBD that is present in the fresh cannabis plant (particularly in its industrial hemp forms). It slowly decarboxylases (that is, loses its acidic function) in response to heating (e.g., when marijuana is smoked). Recent evidence indicates that CBDA is 100–1,000 times as potent as CBD in reducing toxin-induced vomiting and nausea in animal models of chemotherapy treatment. It may be particularly effective in treating the side effect of anticipatory nausea (for which no selective treatment is currently available) in chemotherapy patients (Bolognini et al. 2013; Rock, Limebeer, and Parker 2014; Rock, Sticht, and Parker 2014)

In the chapters that follow, the effects of THC and/or CBD on various brain functions will be described. Other cannabinoid constituents may also be found to have therapeutic potential as more investigators access these compounds for scientific investigation.

Methods of Use

Cannabis can be self-administered by smoking, by vaporization, by eating, by oral application of a tinctures, and by topical application of a salve. Evidence suggests that vaporization (heating until volatile active cannabinoids are vaporized) reduces combustion products in the inhalant and may therefore be a safer method than smoking, though some toxins or carcinogens may be present in vapor (Savage et al. 2016). Smoking or vaporization produces more rapid onset of psychoactive effects—within minutes—than other routes of administration; this allows the user to regulate the desired effect by taking more or fewer puffs (called *titration*). "Edibles" (foods and drinks laced with cannabinoids) are increasingly used but are slower in onset (up to 90 minutes) and entail the risk of unpredictable side effects because controlling doses is difficult. Another concern is that children may mistake edibles for non-cannabis-containing treats. Since cannabinoids are highly lipophilic, they can be topically applied in a salve. Applied in that fashion, they have a local or systemic effect; however, it appears that no studies have been done to evaluate the efficacy of this route of administration. Smoking remains the most common route of administration.

The Pharmacokinetics of THC

Most of what we know about the pharmacokinetics of cannabis constituents has been determined by investigation of THC. Pharmacokinetics assesses the absorption of THC after its administration by a variety of routes; its distribution throughout the body; its metabolism by tissues and organs; its elimination in feces, urine, sweat, and hair; and how these processes change over time.

Bioavailability

The term *bioavailability* refers to the fraction of an administered dose of unchanged drug that reaches the systemic circulation. By definition, when a drug is administered intravenously, its bioavailability is 100 percent. When the drug is administered by smoking or orally, its bioavailability decreases (owing to incomplete absorption or by first pass metabolism by enzymes in the liver); it also may vary from person to person. The pharmacokinetics of THC varies with the route of administration.

Smoking—the most common route of cannabis administration—provides rapid and efficient delivery from the lungs to the brain. The bioavailability of smoked THC is approximately 25 percent (with large variability due to many factors, including the experience of the smoker). Smoked THC reaches peak plasma concentration (C_{max}) in about 6–10 minutes. The mean concentration is approximately 60 percent of the peak concentrations 15 minutes after initiation of smoking and approximately 20 percent 30 minutes after initiation of smoking. Smoking allows the user to titrate the dose. THC metabolizes to an equipotent 11-hydroxy-THC metabolite (11-OH-THC) and an inactive 11-nor-9-carboxy-THC (THCCOOH) metabolite during cannabis smoking (Huestis and Smith 2014).

When cannabinoids are ingested, absorption is slower and peak concentrations are delayed and are lower than when they are smoked. Earlier studies of THC bioavailability found that bioavailability was lower (about 6 percent) after oral ingestion than after smoking (about 25 percent), but administering THC in sesame oil improves bioavailability (Huestis 2005). Time to plasma THC C_{max} is about 2–6 hours after oral administration and about 6–10 minutes after smoking. After a 20-milligram (mg) dose in food (chocolate cookies), peak plasma THC concentrations are 0.004-0.011 mg per liter 1–5 hours after ingestion. Similar concentrations occur after administration of 10 mg of Marinol (dronabinol) in a capsule (Huestis and Smith 2014).

Nabiximols (Sativex; GW Pharmaceuticals, Salisbury, England) was developed for treatment of pain and spasticity due to multiple sclerosis. Nabiximols consists of an approximately 50/50 mixture of Δ^9-THC/CBD (2.7 mg THC: 2.5 mg CBD/spray) and is administered sub-lingually as a spray. A randomized, controlled, double-blind study compared the levels of THC, CBD, 11-hydroxy-THC, and THCCOOH in the plasma from 0 to 10½ hours among occasional marijuana smokers given low (5.4 mg THC and 5.0 mg CBD) and high (16.2 mg THC and 15 mg CBD) doses of oromucosal nabiximols with 5 and 15 mg of synthetic oral THC (Karschner, Darwin, Goodwin, Wright, and Huestis 2011; Karschner, Darwin, McMahon, et al. 2011). Similar bioavailabilities of nabiximols and oral THC were observed, and the addition of CBD to the nabiximols did not modify the bioavailability of THC.

Distribution

The concentration of THC in the blood decreases rapidly after smoking as a result of tissue distribution, hepatic metabolism, and urinary and fecal excretion. Because THC is highly lipophilic, it is rapidly taken up by tissues with high blood flow, including the heart, the lungs, the brain, and the liver. The minimum dose of THC necessary to produce pharmacological effects in humans is between 2 and 22 mg (Huestis 2005). With a bioavailability of 10–25 percent, this means that between 0.2 and 4.4 mg of THC is the actual smoked dose received, with about 1 percent (0.002–0.044 mg) in the brain at peak concentration. Tissues with less blood flow accumulate THC more slowly and release it over a longer period of time; indeed, THC stored in fat in chronic frequent cannabis smokers can be released into the blood for days (Huestis and Smith 2014). It is not entirely certain if THC persists in the brain in the long term, but the presence of residual cognitive deficits in heavy cannabis users raises the possibility that THC may be retained in the brain, at least in the short term (Pope 2002). In an evaluation of THC elimination from the blood in 30 male daily cannabis users who abstained for 33 days and were maintained in a closed residence, both THC and its inactive metabolite, THC-COOH, were detected in blood as long as a month after smoking—four times as long as was previously reported (Bergamaschi et al. 2013).

Metabolism

Most THC metabolism occurs in the liver, and different metabolites are predominant depending upon the route of administration. THC is metabolized in the liver by hepatic cytochrome P450, oxidases 2C9, 2C19 and 3A4 enzymes primarily producing the metabolites 11-OH-THC, THC-COOH and glucuronide conjugates. Plasma concentrations of 11-OH-THC (which has psychoactive properties equal to or greater than those of THC) after smoking are about 10 percent those of THC and peak about 15 minutes after the start of smoking. After oral ingestion, they are about three times as high as after smoking, and they peak after 2–4 hours. The inactive metabolite THC-COOH does not have psychoactive properties. There is a large degree of intra-individual and inter-individual variability in the concentration profiles of plasma THC and metabolites, and there are no significant differences between men and women (Huestis and Smith 2014). CYP isozyme polymorphisms may affect the pharmacokinetics of THC; people homozygous for the CYP2C9*3 allelic variant display significantly higher plasma maximum plasma concentrations of THC and significantly decreased clearance relative to people with the CYP2C9*1 homozygote (Sachse-Seeboth et al. 2009).

Elimination

Within five days, 80–90 percent of a THC dose is excreted, primarily as metabolites, more than 65 percent it in feces and 25 percent of it in urine. Detection times in urine after smoking a 3.55 percent THC cigarette range from two to five days for occasional cannabis smokers but can extend to weeks in chronic daily cannabis smokers (Huestis and Smith 2014). The estimates of the terminal half-life of THC in humans have increased progressively as analytical methods have become more sensitive.

Tolerance

Tolerance is a state of adaptation in which exposure to a drug causes changes that result in a diminution of one or more of the drug's effects over time. Tolerance results from pharmacodynamics or pharmacokinetic mechanisms. Pharmacokinetic (metabolic) tolerance occurs because a decreased quantity of the substance reaches its site of action, which may be caused by an increase in the degradation of the drug by enzymes in the liver. Pharmacodynamic tolerance occurs when the cellular response to a drug is reduced with repeated use, which may be caused by a reduced receptor response (receptor desensitization) or by a reduced density of receptors (receptor downregulation). Pharmacodynamic tolerance to a receptor antagonist produces the opposite: increased receptor density (receptor upregulation) or sensitivity (receptor sensitization). In addition, considerable evidence indicates that tolerance may develop to the effect of a drug by means of a Pavlovian association between the contextual cues in which the drug is taken and the effects of the drug (Siegel 2016); however, there have been no investigations of the role of conditioning in the tolerance to the effects of cannabis.

Tolerance to the effects of THC appears to result primarily from pharmacodynamics rather than from pharmacokinetic mechanisms (Gonzalez, Cebeira, and Fernandez-Ruiz 2005). Such tolerance results when chronic activation of the CB_1 receptor produces either receptor desensitization (uncoupling of the receptor from intra-cellular downstream signal transduction events) or receptor downregulation (resulting from internalization or degradation of the receptor) (Lazenka, Selley, and Sim-Selley 2013). This may explain why tolerance develops to some effects of cannabis but not to others.

Dosing

Cannabis does not fit well with the typical medical model for prescribing drugs. The complex pharmacology of cannabinoids, human genetic differences in metabolism affecting cannabinoid bioavailability, previous experience with recreational cannabis, tolerance, variable potency of the cannabis plant material, and different routes of administration all contribute to the difficulty of reporting precise doses for various conditions. Besides smoking and vaporization, cannabis is consumed in baked goods such as cookies or brownies, or is drunk as teas. Absorption of these products by the oral route is slow, the onset of effects is delayed, and the effects last much longer than with smoking. Dosages for oral administration are even less well established than for smoking or vaporizing (Wachtel, ElSohly, Ross, Ambre, and de Wit 2002). In general, dosing remains highly individualized and relies to a great extent on self-titration. Patients not experienced with cannabis who are undergoing cannabis therapy for the first time are cautioned to begin at a very low dose and to stop therapy if undesirable or unacceptable side effects occur. Consumption of smoked, inhaled, or orally administered cannabis should begin slowly, within a minimum of 10–20 minutes between puffs or inhalations and a minimum of 30 minutes (but preferably 3 hours) between bites of cannabis-based oral products to determine the strength of the effects and whether overdosing has occurred. (For more on dosing, see the Marihuana for Medical Purposes Regulations Daily Amount Fact Sheet on Dosage at the Health Canada website: hc-sc.gc.ca.)

A number of surveys have suggested that most people using smoked or orally ingested cannabis for medical purposes reported using between 10 and 20 grams of cannabis per week. (See, e.g., Ware and Tawfik 2005.) An international Web-based cross-sectional survey of about 950 self-selected participants from a wide variety of countries indicated that the vast majority preferred inhalation over other forms of administration for symptoms such as chronic pain, anxiety, loss of appetite, depression, and sleep disorders. Median daily doses were 2 g of smoked cannabis and 1.5 g of vaporized cannabis (Hazekamp, Ware, Muller-Vahl, Abrams, and Grotenhermen 2013). With orally administered cannabinoids in foods the median daily dose was 1.5 g; with teas the median daily dose was 1.5 g. However, information on cannabis potencies (THC/CBD levels) was not available in this study. The average frequency of use was six times per day for smoking, five times per day for vaporizing, and twice per day for cannabinoids in teas or food. First onset of effects occurred 7 minutes after smoking, 6½ minutes after vaporizing, 29 minutes after ingestion of tea, and 46 minutes after ingestion of food. Satisfaction ratings for ease of finding the correct dose and for lack of side effects were reported to be higher for smoking and vaporizing than for foods or teas. The majority of the survey participants reported having used cannabis before onset of their medical conditions (Hazekamp et al. 2013).

A typical marijuana cigarette contains between 0.5 and 1 g of cannabis plant material, which may vary in THC content between 7.5 and 225 mg (between 7 percent and 30 percent) and in CBD content between 0 and 180 mg (between 0 and 24 percent). The actual amount of THC delivered in the smoke varies, the estimates ranging from 20 percent to 70 percent (Carter, Weydt, Kyashna-Tocha, and Abrams 2004). The actual dose of THC absorbed has been estimated to be about 25 percent of the total amount available in the cigarette. Although cannabis smokers titrate their doses of THC by inhaling lower volumes of high-THC-content cigarettes, doing so may not completely compensate for the higher THC doses per cigarette when using very strong cannabis, resulting in greater exposure to THC (van der Pol et al. 2014).

There has been no documented evidence of a death that can be exclusively attributed to overdosing with cannabis probably because of the sparsity of CB_1 receptors in the brainstem region that controls respiratory and cardiovascular systems (Herkenham et al. 1990). The rare acute complications (e.g., panic attacks, severe anxiety, psychosis, paranoia, convulsions, hyperemesis) that are seen in hospitals' emergency departments are an extension of the psychoactive effects of potent THC or synthetic cannabinoids. As was noted above, inhalation is associated with a large and rapid increase in blood cannabinoid levels, but a short duration of action. On the other hand, oral ingestion results in a delayed onset of action and a longer duration of action. The sudden increase in blood levels of THC or synthetic cannabinoids due to inhalation may produce these adverse side effects if self-titration is not properly employed. There is some evidence that participants do not fully compensate for higher THC doses per cigarette when using strong cannabis (van der Pol et al. 2014). The slow onset of edibles can lead some users to eat more, believing that they have not consumed enough, which also can lead to a dose that produces the adverse effects.

Estimating equivalent human doses on the basis of findings from pre-clinical animal studies is challenging, because most pre-clinical studies administer cannabis constituents (e.g., THC or CBD) by injection or by oral gavage in isolation rather than the composite cannabinoids that humans take when they smoke or eat cannabis. One method that has been used to compare equivalent doses is to administer THC orally by giving marinol to humans and by gavaging rats or mice. The maximal dose of marinol administered to humans is 20 mg/day. This dose taken by a 60-kg human translates to 0.3 mg/kg/day. However, the extrapolation of the human dose to an animal dose cannot be done by a simple conversion based on body weight. In translating human to animal doses, a method of normalizing the surface area of the body is used (Reagan-Shaw et al. 2007); indeed, that is the standard method for converting doses of new drugs discovered in pre-clinical research to doses to be tested in phase I and phase II clinical trials with humans. With such a conversion, the dose equivalent for a 0.02-kg mouse is 12 times higher (e.g., 3.6 mg/kg/day oral marinol), that for a 0.15-kg rat is 6 times higher (1.8 mg/kg/day oral marinol), that for a 3-kg monkey is 3 times higher (0.9 mg/kg/day oral marinol), and that for a 20-kg child is 1.5 times higher (0.6 mg/kg/day oral marinol) than that for an adult human.

This method correlates well across several mammalian species for several biological parameters, including oxygen utilization, caloric expenditure, basal metabolism, blood volume, circulating plasma proteins, and renal function, all of which can effect the activity of drugs.

Conclusion

Although cannabis is one of the oldest known drugs and has been used for many purposes throughout history, not until the early 1990s did we begin to understand how it acts on the brain and the body. The scientific investigation of this endocannabinoid system will lead to a much richer understanding of the effects of this drug and of the medicinal opportunities provided by the constituents of the cannabis plant.

In the twentieth century, laws against possession of cannabis made research on its effects very difficult, particularly in academic institutions. However, since the discovery of the active ingredient in the plant in 1964 (Gaoni and Mechoulam 1964a), the subsequent discovery of the mechanism of action of THC (Devane et al. 1988) and the discovery of the natural ligands that act on cannabinoid receptors (Devane et al. 1992; Mechoulam et al. 1995), there has been a surge of basic *in vitro* and *in vivo* pre-clinical research with animal models on plant cannabinoids and the endogenous cannabinoids. Indeed, the burst of recreational marijuana use in the 1960s in the United States and in Europe coincided with the increase in research on cannabis in academic institutions (Mechoulam and Parker 2013). Some of these new findings will be presented below.

2

The Endocannabinoid System

For nearly a hundred years, investigators attempted to isolate the active constituents of marijuana and to describe their structure, but most of the attempts were unsuccessful (Mechoulam and Hanus 2000). At the end of the nineteenth century, a group at Cambridge (Wood, Spivey, and Easterfield 1896) distilled from Indian charas a red oil extract that produced marijuana-like effects in a research assistant who consumed it (Marshall 1897). Wood's group isolated from the red oil a crystalline compound which they named *cannabinol*. The structure of cannabinol was elucidated by Cahn and Todd in the 1930s in the United Kingdom and by Adams in the United States; Adams also isolated cannabidiol (CBD) from the cannabis plant (Mechoulam et al. 2014). Thus, cannabinol was the first natural cannabinoid to be extracted from cannabis in pure form; however, in the first half of the twentieth century, investigation of cannabinol revealed that it was not the primary active ingredient in cannabis. The search continued for the compound in the cannabis plant that produced the known psychoactive effects of marijuana. Most of these investigations were unsuccessful. As we now know, cannabis has more than 100 constituents, with closely related structures and physical properties. However, in the 1960s the development of modern chromatography separation techniques allowed Gaoni and Mechoulam (1964a) to isolate and identify the active constituent, Δ^9-tetrahydrocannabinol (THC). This allowed THC to be synthesized (Mechoulam and Gaoni 1967), which made it widely available for research. Since then, several thousand papers describing its effects have been published.

After the discovery of THC, it was assumed, because of its lipid chemistry, that it acted through a non-specific membrane-associated mechanism. However, some of the synthetic cannabinoids that were developed had very high stereospecificity, which suggested a specific mechanism of action (Mechoulam et al. 1988). For instance, one very potent synthetic cannabinoid, HU-210, was several thousand times as potent in tests with animals as its synthetic mirror image, HU-211. These results prompted Allyn Howlett's research group in St. Louis to look for a specific cannabinoid receptor in the rat brain. After they found it, they called it a *cannabinoid-1 (CB$_1$) receptor* (Devane et al. 1988). Their data also demonstrated that cannabinoids inhibit adenylate cyclase formation, suggesting that the CB$_1$ receptor was a G-protein-coupled receptor (GPCR) and that the pharmacological action of cannabinoids varied as a function of their potency (Herkenham et al. 1990). Later, a second, peripheral receptor, CB$_2$, was identified in the spleen (Howlett et al. 2002; Munro et al. 1993). Both CB$_1$ and CB$_2$ receptors belong to the superfamily of GPCRs. Both receptor types are coupled through G proteins to adenylyl cyclase and mitogen-activated protein kinase. (For a detailed review of the pharmacology of cannabinoids, see Mechoulam et al. 2014. For a glossary of pharmacological terms used in the present chapter, see table 2.1.)

The CB$_1$ Receptor

The CB$_1$ receptor was initially believed to be expressed mainly in the central nervous system (CNS), and hence it was considered a brain cannabinoid receptor. However, it is now clear that it is also present in some peripheral organs (e.g., heart, liver, fat tissue, stomach, testes) at a lower expression level than in the brain. In the brain, CB$_1$ receptors are among the most abundant GPCRs. The highest densities of CB$_1$ receptors in the rodent brain are found in the basal ganglia, in the substantia nigra, in the globus pallidus, in the cerebellum, and in the hippocampus, not in the brainstem. (See figure 2.1.) Since the brainstem regulates basic autonomic functions necessary for survival (e.g., breathing), the paucity of CB$_1$ receptors in it explains the low toxicity of Δ^9-THC. The high concentration of CB$_1$ receptors in the sensory and motor regions is consistent with the importance of CB$_1$ receptors to motivation and cognition. CB$_1$ receptors appear to be involved in γ-aminobutyric acid (GABA) and glutamate neurotransmission, as they are found on GABAergic and glutamatergic neurons (Howlett et al. 2002). CB$_1$ receptors are also found in the peripheral nervous system on sensory neurons, on postganglionic sympathetic neurons, and on neurons of the parasympathetic nervous system. The gut (enteric) nervous system is also rich in CB$_1$ receptors (Sharkey, Darmani, and Parker 2014).

Table 2.1
Glossary of pharmacological terms.

Affinity	The potency with which a compound binds to a particular receptor. The higher the affinity, the lower the concentration at which it achieves a specified level of receptor occupancy.
Agonist	A compound that activates a receptor. A full agonist is more potent than a partial agonist and produces a greater maximal functional response at the receptor.
Allosteric modulator	A substance that acts on an allosteric site of a receptor to increase or decrease the ability of the natural ligand to induce a functional response when it binds to the orthosteric site on the same receptor.
Antagonist	A compound that binds to a receptor by targeting its orthosteric site but does not activate the receptor. An antagonist can prevent both drug-induced and natural-ligand-induced agonism of the receptor.
Efficacy (E_{max})	The maximum response achievable from a drug or a natural ligand for the receptor
Functional selectivity	Different agonists can activate different suites of signaling pathways at the same receptor. Functional selectivity is also called *agonist trafficking* or *biased agonism*.
G-protein-coupled receptor (GPCR)	A seven-transmembrane domain receptor that induces G-protein-mediated activation of intracellular signal transduction pathways when occupied by an agonist.
Inverse agonist	A compound that binds to a receptor in a manner that induces a pharmacological response opposite to the response induced by an agonist for the same receptor.
Receptor desensitization	Chronic exposure to receptor agonists decreases the signaling response of the receptor.
Receptor downregulation	Chronic exposure to receptor agonists temporarily decrease the number of receptors.
Receptor upregulation	Chronic exposure to receptor antagonists temporarily increase the number of receptors.
Relative intrinsic activity	The ability of drug-receptor complexes to produce maximum functional responses. A full agonist needs to occupy fewer receptors than a partial agonist to produce a maximal response.
Retrograde synaptic messenger	A compound that is released by a post-synaptic dendrite or cell body, but that acts pre-synaptically.

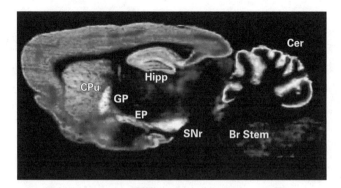

Figure 2.1
Distribution of brain CB1 receptors shown by autoradiography of CP 55,940 (potent CB1 agonist) binding in the rat brain. Gray levels represent relative levels of receptor densities. Saggital section of rat brain. Br Stem indicates brainstem; Cer indicates cerebellum; CP indicates caudate-putamen; Cx indicates cerebral cortex; Ep indicates entopeduncular nucleus; GP indicates globus pallidus; Hipp indicates hippocampus; SNr indicates substantia nigra. Based on Herkenham 1991.

The CB$_1$ receptors are found primarily on central and peripheral neurons in the pre-synapse, as shown in figure 2.2. The most frequently reported neuronal effect of CB$_1$-receptor agonists is inhibition of synaptic transmission. These locations facilitate their inhibition of neurotransmitter release, which is one of the major functions of the endocannabinoid system. Neurons release their neurotransmitter substance when an action potential arrives at their pre-synaptic terminal ending. The action potential results in an influx of Calcium (Ca^{2+}) which stimulates the release of the neurotransmitter. (In figure 2.2, the neurotransmitter is glutamate, the primary excitatory neurotransmitter in the brain.) However, activation of the pre-synaptic CB$_1$ receptor leads to a decrease in cyclic adenosine monophosphate (cAMP) accumulation, which inhibits the influx of Ca^{2+} in the firing neuron and hence inhibits neurotransmitter release.

Endogenous Cannabinoid Agonists

The discovery of the cannabinoid receptors suggested that endogenous molecules, which may stimulate (or inhibit) the receptors, might be present in the mammalian body. Earlier in the 1970s the discovery of the opiate receptor in the brain prompted an intense search for the naturally occurring brain chemicals that act on this receptor; that search revealed the family of peptides known as endorphins (endogenous morphines).

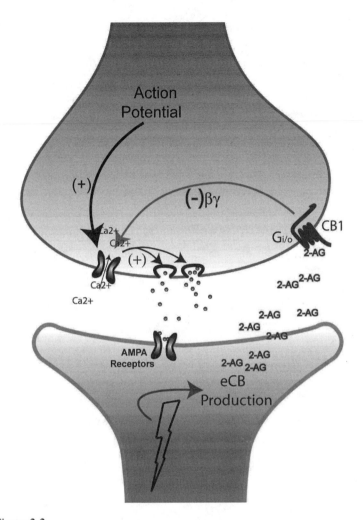

Figure 2.2

A synaptic junction. Upon an action potential arriving at the pre-synaptic terminal ending, glutamate is released into the synapse. Glutamate acts on AMPA receptors on the adjacent dendrite resulting in depolarization of the post-synaptic membrane. The subsequent influx of Ca++ results in the production of 2-AG, which is released into the synapse and serves as a retrograde messenger acting on inhibitory pre-synaptic CB1 receptors on the glutamate-containing neuron to turn off subsequent glutamate release. The action of 2-AG is terminated by its degrading enzyme, monoacylglycerol lipase (MAGL), located in the pre-synaptic neuron.

Similarly, the discovery of cannabinoid receptors prompted a search for naturally occurring cannabinoids. The plant constituent THC, which just happens to bind to these receptors, is a lipid compound; hence it was assumed that any possible endogenous cannabinoid molecules (endocannabinoids) would also be lipids. Mechoulam and his colleagues isolated and identified two such compounds—one from brain tissue (anandamide, abbreviated AEA, its name derived from the Sanskrit word *ananda*, meaning "supreme joy"; see Devane et al. 1992) and one from peripheral tissues (2-arachidonoyl glycerol, abbreviated 2-AG; see Mechoulam et al. 1995). The latter was also isolated from brain tissue (Sugiura et al. 1995) and is therefore found in both the periphery and in central systems. Thousands of articles on these two compounds have been published; for a review, see Mechoulam et al. 2014. Their discovery has revealed a new, unique biological system of control: the *endocannabinoid system*. Additional endogenous molecules that bind to the cannabinoid receptors (virodhamine and noladin ether) have been identified, but much less research has been done with them.

In contrast with most neurotransmitters (e.g., glutamate, GABA, acetylcholine, dopamine, and serotonin), AEA and 2-AG are not stored in vesicles but rather are synthesized on demand when and where they are needed. The action of both AEA and 2-AG on the CB_1 receptor is

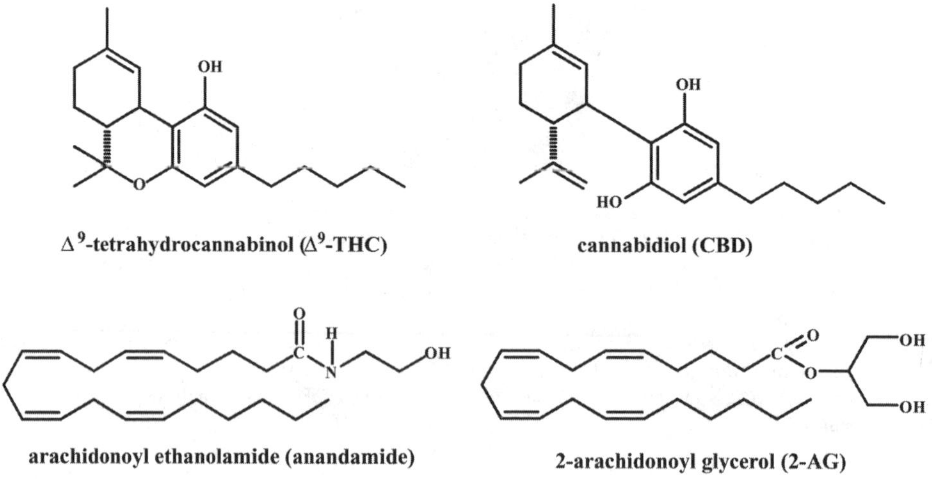

Δ^9-tetrahydrocannabinol (Δ^9-THC)

cannabidiol (CBD)

arachidonoyl ethanolamide (anandamide)

2-arachidonoyl glycerol (2-AG)

Figure 2.3
Molecular structures of THC, CBD, AEA, and 2-AG.

primarily pre-synaptic, not post-synaptic, and their main function is to reduce the neuronal release of other transmitters (e.g., glutamate). It is clear that 2-AG, in particular, acts as a fast retrograde synaptic messenger (Wilson and Nichol 2001) to regulate neurotransmission in the brain. (See figure 2.2.) The synthesis of 2-AG is triggered by the increase in the concentration of intracellular calcium when the post-synaptic neuron becomes depolarized by the action of a neurotransmitter (e.g., glutamate) on excitatory post-synaptic receptors. Therefore, it is produced only in regions that are activated. The resulting influx of calcium induces post-synaptic biosynthesis and a release of 2-AG into the synapse. 2-AG is formed from diacylglycerol (DAG) in a process that is catalyzed by sn1-specific DAG lipase-α. The main biosynthetic pathway for AEA involves the formation of N-arachidonlyl phosphatidylethanolamine (NAPE) from phosphatidylethanolamine and phosphatidylcholine. NAPE is then converted to AEA by the action of the enzyme NAPE selective phospholipase D. However, the on-demand eCB synthesis of AEA is not as well understood as that of 2-AG (Mechoulam et al. 2014).

Whereas THC is metabolized over several hours and excreted (or stored as one of its metabolites), endocannabinoids are rapidly removed by a membrane transport process yet to be fully characterized (Fu et al. 2012; O'Brien et al. 2013; Leung et al. 2013). Once AEA crosses the post-synaptic membrane, fatty-acid binding proteins (Kaczocha, Glaser, and Deutsch 2009) transport AEA to intracellular fatty-acid amide hydrolase (FAAH) for inactivation. FAAH hydrolyzes AEA to arachidonic acid and ethanolamine (Cravatt et al. 1996). In the pre-synaptic cell, 2-AG is also hydrolyzed enzymatically, by monoacylglycerol lipase (MAGL) (Dinh, Kathuria, and Piomelli 2004). Suppression of these enzymes prolongs the activity of the endocannabinoids. In addition to these two primary metabolic enzymes, AEA and 2-AG, are also oxygenated by cyclooxygenase 2 (COX-2) to form bioactive prostaglandin derivatives (Hermanson et al. 2013). And a small proportion of 2-AG is also metabolized by alpha-beta hydrolase AβHD class of enzymes, specifically AβHD6 and AβHD12 (Blankman, Simon, and Cravatt 2007; Marrs et al. 2010). The endocannabinoid receptors, the endocannabinoids, and their biosynthetic and biodegrading enzymes constitute what is known as the *endocannabinoid system*.

In general, activation of the endocannabinoid system at the synapse leads to a short or sustained suppression of neurotransmitter release from the pre-synaptic terminal afferent. Although 2-AG and AEA probably act in a similar manner to regulate transmitter release, it is believed that AEA acts to regulate tonic basal synaptic transmission, whereas 2-AG acts as the phasic signal triggered during sustained neuronal depolarization and mediates many forms of synaptic plasticity (Ahn et al. 2009; Katona and Freund 2012). In addition, 2-AG is a full agonist (i.e., produces a greater maximal effect) of both the CB_1 and CB_2 receptors, whereas AEA is a partial agonist, as is THC. The concentration of 2-AG in the brain is higher, and 2-AG is more broadly expressed than AEA; indeed, 2-AG acts as the primary working endocannabinoid in the brain (Katona and Freund 2012). AEA appears to be activated when the physiological systems are stressed (Hill et al. 2010).

Although it is clearly understood that activation of pre-synaptic CB_1 receptors can lead to inhibition of the release of a number of different excitatory or inhibitory neurotransmitters both in the brain and in the peripheral nervous system, there is also *in vivo* evidence that CB_1-receptor agonists (Nunez et al. 2004) can stimulate dopamine release in the nucleus accumbens (Gardner 2005). This effect apparently stems from a cannabinoid-receptor-mediated inhibition of GABA release that indirectly elevates dopamine release. (See figure 6.3.) Many of the actions of cannabinoid-receptor agonists (including endocannabinoids) are dose-dependently biphasic (Sulcova, Mechoulam, and Fride 1998). (For a review of the multiple functions of endocannabinoid signaling in the brain, see Katona and Freund 2012.)

The CB_2 Receptor

It had been assumed that CB_2 receptors were present only in cells of the immune system (such as β-lymphocytes, monocytes, macrophages, mast cells, and microglial cells); however, they have now been identified in the CNS (Onaivi et al. 2008; Van Sickle et al. 2005), particularly in microglial cells (Nunez et al. 2004; Stella 2004), though at lower levels than those of the CB_1 receptors. Under some pathological conditions, CB_2-receptor expression is enhanced in the CNS as well as in other tissues. However, whether CB_2 receptors are present in a non-injured brain is a matter of debate. It has been suggested that the CB_2 receptor is part of a general protective system. (For a review, see Pacher and Mechoulam 2011.) "The mammalian body," Pacher and Mechoulam speculated, "has a highly developed immune system which guards against continuous invading protein attacks and aims at preventing, attenuating or repairing the inflicted damage. It is conceivable that analogous biological systems that protect against non-protein attacks have evolved. Evidence is emerging that lipid endocannabinoid signaling through CB_2 receptors may represent an example/part of such a protective system" (ibid., p. 194). Several synthetic CB_2-specific receptor agonists (e.g., HU-308; see Hanus et al. 1999) have been synthesized to determine their neuroprotective potential. Since they do not cause the psychoactive effects associated with CB_1 agonists, CB_2-receptor agonists might be expected to become drugs in various fields, including pain, cardiovascular disease, and liver disease.

Development of Drugs That Inhibit FAAH or MAGL

Systemic administration of FAAH inhibitors that penetrate the brain produces increases in central AEA (and other fatty acids) without affecting 2-AG levels, and the opposite is seen for selective MAGL inhibitors (Ahn et al. 2009; Ahn et al. 2011; Kathuria et al. 2003). The rationale behind this approach to drug development is based on the mechanism of AEA or 2-AG formation and release, which is known to take place when and where needed (on demand). As has been noted, AEA and 2-AG are not stored in synaptic vesicles but rather are synthesized and released in the synaptic cleft after neuronal activation. Therefore, inhibition of AEA or 2-AG metabolism would enhance CB_1 activation mainly where AEA or 2-AG levels are highest, but not globally throughout the brain/body. Such an approach produces elevated levels of endocannabinoids in these regions for as long as 24 hours.

One advantage of the using treatments that boost the natural endocannabinoid system is that these treatments do not produce the same psychoactive effects of global CB_1 agonists. FAAH inhibitors do not produce the typical psychoactive effects of THC, as assessed in the "mouse tetrad" assay. They do not produce hypokinesis (sedation), catalepsy, or hypothermia (Ahn et al. 2011; Kathuria et al. 2003); although they do produce analgesia effects. In humans, FAAH inhibitors did not show any effects on cognitive function, psychomotor function, attention and learning after 1, 8 or 14 days of treatment (Li, Winter, et al. 2012). MAGL inhibitors also do not produce catalepsy or hypothermia, unlike THC; however, they have been reported to produce a reduction in locomotor activity and similar degree of analgesia effect as FAAH inhibition (Long, Nomura, and Cravatt 2009). On the other hand, the more selective compounds KML29 and MJN110 have been reported to lack the sedative properties of earlier MAGL inhibitors (Ignatowska-Jankowska et al. 2014; Niphakis et al. 2013; Parker, Niphakis, et al. 2015). FAAH and MAGL inhibitors are being developed and tested for a number of conditions, including pain, inflammation, and cannabis withdrawal disorder. However, considerable pre-clinical animal evidence indicates their potential in the treatment of anxiety, depression, post-traumatic stress disorder, appetite enhancement, addiction, epilepsy, neurodegenerative disorders, and nausea.

Transient Receptor Potential Cation Channel Subfamily V Member 1 (TRPV1)

AEA not only acts on CB_1 and CB_2 receptors; it also acts on TRPV1 receptors. When AEA is produced post-synaptically at the metabotropic glutamate receptor 5 ($mGLU_5$), it also activates post-synaptic intracellular TRPV1 receptors (Alger 2012), which are activated by capsaicin (the hot stuff in chili peppers) and can suppress pain sensitivity (Gregg et al. 2012). AEA interacts with TRPV1 at the same intracellular binding site as capsaicin. The binding site is located on the inner face of the plasma membrane, which suggests that when AEA is biosynthesized by cells that express TRPV1 it will activate that receptor before being released into the synapse, thereby regulating Ca^{2+} homeostasis as an intracellular messenger (van der Stelt and Di Marzo 2005). Consistent with the hypothesis that it is involved in pain, nociception, and temperature sensitivity, TRPV1 is predominantly expressed in sensory neurons of the dorsal root of the spinal cord and the trigeminal ganglion, where it can become upregulated during nerve injury and diabetic neuropathy (Pertwee 2009). Evidence has also accumulated that TRPV1 is found in brain neurons and in non-neural cells. TRPV1 colocalizes with CB_1 receptors in sensory and brain neurons and with CB_2 receptors in sensory neurons. This colocalization makes possible several types of intracellular cross talk between ligands that activate both types of receptors. It is now known that AEA, but not 2-AG, binds to both human and rat TRPV1, upon which it acts as a full agonist. The affinity of AEA to TRPV1 is slightly lower than that of capsaicin to TRPV1 (Di Marzo, Blumberg, and Szallasi 2002), but the activity on these receptors depends upon the assay being used to assess their pharmacological activity. In addition, the affinity of AEA to the CB_1 receptor is higher than its affinity to the TRPV1 receptor; this accounts for the findings that exogenous AEA, by boosting the endogenous AEA signal, can exert effects at TRPV1 more effectively in the presence of an FAAH inhibitor.

THC and synthetic CB_1 and/or CB_2 agonists interact with TRPV1, but usually with a lower relative intrinsic activity and potency than is exhibited by AEA, but N-arachidonoyl dopamine displays a much greater potency than AEA as a TRPV1 agonist. In contrast, CBD and CBG both act as full agonists at TRPV1 receptors (Ligresti et al. 2006). CBD is almost as potent as capsaicin at the human TRPV1 receptor, but is less potent at the rat TRPV1 receptor (Qin et al. 2008). Oleoyl ethanolamide (OEA) is also a potent TRPV1-receptor agonist, but has little or no activity at CB_1/CB_2 receptors

G Protein Receptor 55 (GPR55)

Because some effects of compounds thought to be specific for CB_1 or CB_2 receptors appeared to also act independently of these receptors, a third cannabinoid receptor was suggested to exist (Derocq et al. 1998). That receptor, now known as GPR55, was cloned in 1999 and was identified as a G-protein coupled receptor highly expressed in the human brain (Sawzdargo et al. 1999). There was considerable controversy in the early literature after this receptor was discovered regarding whether AEA or 2-AG activates GPR55 (Liu, Song, Jones, and Persaud 2015). More recently it has become clear that a bioactive lipid, lysophosphatidylinositol (LPI), acts as an agonist on the GPR55 receptor (Oka, Nakajima, Yamashita, Kishimoto, and Sugiura 2007). CBD acts at multiple targets, including acting as an antagonist on the GPR55 receptor. Finally, the CB_1 inverse antagonist, rimonabant, acts as an agonist at the GPR55 receptor, although it requires a higher concentration than is required to antagonize the CB_1 receptor (Pertwee, Thomas, Stevenson, Maor, and Mechoulam 2005; Ryberg et al. 2007). One site of action of GPR55 receptors is on excitatory axon terminals, where they facilitate glutamate release when the neuron fires. Because CBD effectively blocks GPR55 activation, it would be an ideal candidate as an anticonvulsant agent. Thus CBD may reduce excess presynaptic glutamate release from the hyperactive excitatory neurons during epileptic seizures (Sylantyev et al 2013; Katona 2015).

There is compelling evidence suggesting a role for GPR55 in weight gain. Moreno-Navarrete et al. (2012) found that circulating LPI levels were elevated in obese patients and that they correlated positively with body weight, body mass index, and fat percentage in female subjects. Furthermore, GPR55 has been implicated in anorexia nervosa, an eating disorder characterized by excessive food restriction. Ishiguro et al. (2011) found a genetic association between increased vulnerability to anorexia nervosa and dysfunctional alteration of the GPR55 gene at the Val195 allele. The observation that the Val195 allele of GPR55 (which had a reduced signaling capacity) was linked to an increased risk of anorexia nervosa is consistent with inactive GPR55-receptor activity being associated with reductions in food intake. New evidence suggests that the action of LPI on GPR55 facilitates cancer cell proliferation and migration. Indeed, clinical data initially identified LPI as a biomarker for poor prognosis in cancer patients, which is consistent with *in vitro* studies demonstrating elevated levels of LPI in highly proliferative cancer cells.

Peroxisome Proliferator-Activated Receptors (PPARs)

PPARs are ligand activated transcription factors that constitute part of the nuclear receptor family (Pertwee et al. 2010). Classical agonists of PPARs are fatty acids and their derivatives, including oleic acid, arachidonic acid, leukotriene β4, and prostaglandin J2. It is a widespread view that PPARs are not activated by a single endogenous ligand but rather are generalized lipid sensors, monitoring local changes in metabolism. There are three PPAR isoforms: PPARα, PPARβ, and PPARγ. PPARα is a therapeutic target for type 2 diabetes. Oleoyl ethanolamide (OEA) and palmitoyl ethanolamide (PEA), two fatty-acid ethanolamides that are inactive at CB_1/CB_2 receptors, are potent agonists of PPARα. Interestingly, FAAH metabolizes not only AEA but also OEA and PEA; therefore, administration of FAAH inhibitors elevates AEA, OEA, and PEA. A necessary step in determining the mechanism of action of administration of FAAH inhibitors, therefore, is to block the effects with a CB_1 antagonist (then attributing effect to AEA) or with a PPARα antagonist (then attributing effect to OEA and or PEA) (Pertwee et al. 2010).

Allosteric Binding Sites of the CB_1 Receptor

The CB_1 receptor has two binding sites: the main pocket (called the *orthosteric* site) and a secondary one (called the *allosteric* site). A molecule that binds to the allosteric site changes the shape of the main receptor. This change in shape alters the intensity of the effect of any molecule that binds to the orthosteric site (Price et al. 2005). However, allosteric modulators that bind to these allosteric sites have no physiological effect in the absence of the agonism of the orthosteric site. Since the discovery of the allosteric binding site on the CB_1 receptor, compounds have been synthesized that act either as agonists at the allosteric site (they are called positive allosteric modulators) or as antagonists at the allosteric site (negative allosteric modulators). Positive allosteric modulators enhance CB_1-receptor activity, whereas negative allosteric modulators act to antagonize CB_1-receptor activity, both without effects on their own.

An exciting new finding has revealed that the body makes both positive (lypoxin A_4 [Pamplona et al. 2012] and negative (pepcan-12 [Bauer et al. 2012] and pregnenolone [Vallee et al. 2014]) endogenous allosteric modulators that may serve to fine-tune the effects of AEA and 2-AG when released. It is particularly interesting that, although the neurosteroid hormone pregnenolone is normally undetectable in the brains of rat and mice, THC substantially increases the synthesis of pregnenolone in the brain (particularly in the nucleus accumbens, prefrontal cortex, striatum, and thalamus) via activation of the CB_1 receptor. Pregnenolone then acts as a negative allosteric modulator of the CB_1 receptor to reduce several effects of THC in a negative feedback loop, protecting the brain from CB_1 over-activation (Vallee et al. 2014). This exciting finding suggests that pregnenolone may be an effective treatment for cannabis intoxication and addiction. However, there have been some inconsistent findings about the potential of lypoxin A_4 and pregnenolone to modulate 2-AG mediated synaptic transmission (Stricker et al. 2015). There clearly is a need for further research to identify the physiological outcomes of allosteric modulation by endogenously released pepcan-12, lypoxin A_4, and pregnanolone (Pertwee 2015).

Endogenous allosteric compounds may fine-tune the effects of phytocannabinoid drugs, potentially reducing or enhancing unwanted side effects. Indeed, cannabis itself may contain compounds that act as allosteric modulators of the CB_1 receptor or the CB_2 receptor. Recent evidence suggests that CBD behaves as a negative allosteric modulator of the effects of THC and 2-AG (Laprarie et al. 2015). Allosteric modulation at the CB_1 receptor offers new opportunities for therapeutic applications that may fine-tune the signaling pathways of the CB_1 receptor (Morales et al. 2016).

The Role of the Endocannabinoid System in the Development of the Nervous System

The CB_1 receptor is highly expressed in the developing mouse brain (Gaffuri 2012) as early as embryo day 1.5 and into late fetal stages (embryo day 21—the day before birth), with high expression in white matter within a number of different structures including the hippocampus, cerebellum, caudate-putamen, and cerebral cortex. Expression in these regions continues to increase after birth and into adulthood, whereas after birth there is tapering of CB_1 expression in the corpus callosum, fornix, stria terminalis, and fasciculus retroflexus (ibid.). Furthermore, in the adult brain the CB_1 receptor appears to be localized on the axonal plasma membrane and in somatodendritic endosomes, whereas in the fetal brain the CB_1 receptor is mostly localized to endosomes in axons and in the somatodendritic region (ibid.). The evidence suggests that the endocannabinoid system is critical for neuronal development survival, proliferation, migration, and differentiation of neuronal progenitors (ibid.). CB_1-receptor activation in response to stimulation by endocannabinoids, such as 2-AG and AEA, promotes these functions by pruning the proliferation of synapses, contributing to a more settled and better differentiated post-mitotic neuronal phenotype (Zhou 2014; Gaffuri 2012). *In vitro* studies examining the effects of CB_1-receptor activation in primary neuronal cultures suggest that the CB_1 receptor is mainly a negative regulator of neurite growth, since activation of the CB_1 receptor results in arrest, repulsion, or collapse of the growth cone, and that in doing so it influences the ability of axons to reach their targets (Gaffuri 2012). Furthermore, the CB_1 receptor appears to also act as a negative regulator of synaptogenesis and, in doing so, perhaps to affect the fate of neuronal communication (ibid.). Exposure during pregnancy to cannabinoids (such as THC) that activate the CB_1 receptor may alter the course of normal neuronal development in offspring and negatively affect normal brain function, potentially causing long-lasting impairment in a number of cognitive functions and behaviors (Zhou 2014).

Conclusion

The endocannabinoid system, discovered only 20 years ago, is now known to be a major modulator of synaptic activity throughout the brain. CB_1 receptors are most prominently located on the pre-synaptic terminals of neurons that release other neurotransmitters, including excitatory glutamate and inhibitory GABA, as well as acetylcholine (ACh), 5-hydroxytryptamine (5-HT, serotonin), and norepinephrine (NE). At these sites, CB_1 agonists, as well as 2-AG and AEA, act to reduce neurotransmitter release. THC administered through smoking or ingestion of marijuana activates all of the CB_1 receptors to which it binds, producing global activation. On the other hand, endogenously released 2-AG and AEA are produced only where and when they are needed and act to feed back on the pre-synaptic neuron to "turn off" neurotransmitter release. Fine-tuned regulation of synaptic activity is the primary function of this ubiquitous neuromodulatory system that plays a major role in protecting neurons.

The duration of action of these "on-demand" endocannabinoids is brief because they are hydrolyzed enzymatically by FAAH (AEA) and MAGL (2-AG) and removed from the synapse by a re-uptake process that has not yet been characterized. However, the newly developed selective FAAH and MAGL inhibitors provide a therapeutic opportunity to boost the action of AEA and that of 2-AG (respectively) for as long as 24 hours. Results from pre-clinical animal studies have suggested several therapeutic options—including the relief of pain, anxiety, depression, and nausea—for these exciting new medications, which boost the action of endocannabinoids when and where they are needed without the psychoactive side effects of global action of systemically administered CB_1 agonists.

3

Cannabinoids and Emotional Regulation

Among the most prominent effects of cannabis exposure in humans are its effects on emotional behavior, mood, and anxiety; marijuana is often used for relaxation. Therefore, once THC was discovered to be the psychoactive ingredient in cannabis, the earliest experimental investigation of this phytocannabinoid focused on its effects on mood and anxiety. Since the early investigations in the 1970s, our understanding of the mechanism by which THC and other CB_1 agonists affect processes involved in neural regulation of mood and anxiety has advanced considerably. Simply put, activation of the CB_1 receptor is critical for the regulation of the body's systems for signaling stress responses (Hill, McLaughlin, et al. 2010; Hill, Patel, et al. 2010). Recent evidence suggests that treatments that enhance endocannabinoid signaling may be a promising treatment post-traumatic stress disorder (PTSD).

Cannabis and Anxiety

Patients with anxiety disorders report a high incidence of cannabis use, especially in times of stress. This suggests that they are using cannabis to reduce their anxiety symptoms. Although initial use of cannabis can reduce anxiety symptoms, long-term use can worsen anxiety and promote panic attacks. Whether cannabis is used to self-medicate anxiety or whether chronic cannabis use contributes to the disorder is difficult to discern in humans (Patel, Hill, and Hilliard 2014).

THC and anxiety

Laboratory studies with animals support tension reduction with low doses of THC (anxiolytic effects), but chronic administration of high doses of THC produces anxiogenic effects and neuroendocrine responses to acute stress in rats. Even acute administration of higher doses of THC under conditions of high stress increases anxiety in animals by increasing neuronal activity in the limbic system, particularly the amygdala. The higher concentrations of THC found in cannabis in recent years may be related to the increased reports of panic attacks in humans. (See Patel et al. 2014 for an excellent review.)

In pre-clinical studies using new sophisticated genetic tools, advances are being made in understanding the well-known biphasic influence of THC on anxiety-like behavior. The conditional deletion of the gene encoding the CB_1 receptor can be imposed on excitatory glutamate neurons and/or on the inhibitory GABA neurons. The loss of CB_1 receptors in forebrain glutamatergic neurons (which are connected to several regions mediating anxiety behaviors) enhances anxiety-like behaviors in some animal models and is accompanied by elevated glutamate levels in these regions. Conversely, the loss of CB_1 receptors in forebrain GABAergic neurons reduces anxiety-like behaviors and is accompanied by elevated GABA levels in these regions. This opposing control of anxiety-like behaviors by CB_1 receptors on cortical glutamatergic and GABAergic neurons occurs only when the environmental aversiveness exceeds a certain threshold—that is, the endocannabinoid system appears to buffer neuronal activity in specific circuits only in times of stress. By selectively knocking down CB_1 receptors in either GABA or glutamate neurons, investigators have demonstrated that the anxiolytic effect of cannabinoids at low doses depends on CB_1-receptor activation of cortical glutamatergic neurons (thereby decreasing the release of this excitatory neurotransmitter), whereas the anxiogenic effect at high doses is mediated by CB_1-receptor activation of forebrain GABAergic neurons (thereby decreasing the release of this inhibitory neurotransmitter). (See Lutz, Marsicano, Maldonado, and Hillard 2015.)

Cannabidiol and anxiety

Cannabidiol (CBD) is a non-psychoactive cannabinoid that is well tolerated by humans across a wide dose range—up to 1,500 milligrams per day (oral)—with no psychomotor slowing, negative mood effects, or abnormalities in of vital signs (Bergamaschi et al. 2013). It also has a broad pharmacological profile, including interactions with several receptors known to regulate fear and anxiety-related behaviors, including CB_1, $5\text{-}HT_{1A}$, and TRPV1 (Mechoulam, Peters, Murillo-Rodriguez, and Hanus 2007; Pertwee 2009). CBD has been studied in a wide range of animal models of general anxiety. Initial studies of CBD in these models showed that low doses (10 mg/kg) were anxiolytic, but high doses (100 mg/kg) were ineffective (Zuardi, Finkelfarb, Bueno, Musty, and Karniol 1981). Subsequent studies revealed a bell-shaped curve, with anxiolytic effects at moderate doses but not at higher doses (Guimaraes, Chiaretti, Graeff, and Zuardi 1990). There have been no reports of anxiogenic effects of CBD from systemic studies—it is either anxiolytic or without effect at higher doses. Chronic exposure to CBD maintains this anxiolytic effect in these models of anxiety-like behavior.

There is evidence that these anxiolytic effects of CBD are mediated by its action on 5-HT_{1A} receptors in the midbrain dorsal periaqueductal gray (DPAG), a region integral to anxiety that controls autonomic and behavioral responses to threat (Campos and Guimaraes 2008). DPAG stimulation in humans produces feelings of intense distress and dread (Nashold, Wilson, and Slaughter 1969). The bell-shaped dose-response curve of CBD may reflect activation of TRPV1 receptors at the higher dose, as blockade of TRPV1 receptors in the DPAG rendered an ineffective high dose of CBD to be anxiolytic (Campos and Guimaraes 2009). Since activation of TRPV1 receptors has anxiogenic effects, this may indicate that at higher doses, CBD interacts with TRPV1 receptors to prevent the anxiolytic effect, perhaps by acting to elevate endogenous AEA by acting as an FAAH inhibitor. (See, e.g., Mechoulam et al. 2007.) In addition, the bed nucleus of the stria terminalis (BNST) receives input from the central nucleus of the amygdala, and both structures are involved in the expression of anxiety. Infusions of CBD into the BNST and the central amygdala also produce anxiolytic effects, also by a 5-HT_{1A} mechanism of action. CBD has also been shown to reduce the behavioral and cardiac expression of classically conditioned fear elicited by a shock-paired context when injected systemically or directly into the BNST (Gomes et al. 2013; Resstel, Joca, Moreira, Correa, and Guimaraes 2006). As with its effect on natural anxiety, this effect was mediated by its action on the 5-HT_{1A} receptor. As with THC, CBD has been reported to enhance the extinction of contextually conditioned fear responses, which suggested that it may be an effective treatment for PTSD, without the psychoactive side effects of THC. Overall, the existing pre-clinical evidence supports the potential of CBD to serve as an effective treatment for anxiety disorders and PTSD. CBD has a broad range of actions, reducing anxiety, panic, compulsive actions, autonomic arousal, and fear expression. Most of these anti-anxiety effects of CBD can be reversed by blockade of the 5-HT_{1A} receptor, which suggests that this is the mechanism by which CBD produces its anti-anxiety effects.

Human studies confirm the anxiolytic effects of CBD revealed in preclinical animal studies. The anxiolytic effects of CBD were first revealed by studies suggesting that it reversed the anxiogenic effects of THC; CBD (300–600 mg) reduced THC-induced anxiety, but had no effect by itself (Karniol, Shirakawa, Kasinski, Pfeferman, and Carlini 1974; Zuardi, Cosme, Graeff, and Guimaraes 1993). However, CBD potently reduced experimentally induced anxiety (Hindocha et al. 2015; Martin-Santos et al. 2012), including anxiety associated with simulated public speaking in healthy human participants (Zuardi, Shirakawa, Finkelfarb, and Karniol 1982) and in participants with seasonal affective disorder (Bergamaschi, Queiroz, Chagas, et al. 2011). A thorough review of the safety profile (Bergamaschi, Queiroz, Zuardi, and Crippa 2011) verified that cannabidiol is safe and well tolerated in humans.

The potential of THC and CBD as treatments for PTSD

Post-traumatic stress disorder (PTSD) is characterized by re-experiencing a severe traumatic event that causes avoidance of trauma-related cues and hypervigilance. Characteristic symptoms include persistent, intrusive recollections, re-experiencing of the original traumatic events (through dreams, nightmares, and dissociative flashbacks), numbing, avoidance, and increased arousal (Fraser 2009). Sleep disturbance occurs in up to 90 percent of the cases (Jetly, Heber, Fraser, and Boisvert 2015). This disorder is associated with higher use of cannabis than of stimulants, and rates of PTSD are higher in patients with cannabis-use disorder. PTSD patients appear to use cannabis to reduce tension, because lack of improvement after treatment in a residency program for veterans predicts more frequent use of cannabis in the next four months (Bonn-Miller, Vujanovic, and Drescher 2011). Interestingly, the synthetic THC nabilone was shown to reduce the frequency of nightmares and to reduce daytime flashbacks in a subset of veterans, which suggested that cannabis just before sleeping may be a beneficial treatment for PTSD patients (Fraser 2009). More recently, research by Roitman, Mechoulam, Cooper-Kazaz, and Shalev (2014) showed that 5 mg/kg of THC twice a day in addition to the patient's other drugs led to significant improvements in sleep quality, frequency of nightmares, and severity of symptoms overall. However, the sample size was rather small, and there was no placebo control.

A preliminary, randomized, double-blind, placebo-controlled crossover clinical study of ten Canadian male military personnel with PTSD who were not responsive to conventional treatment was recently reported (Jetly et al. 2015). The participants received 0.5 mg nabilone or placebo and titrated to an effective dose (up to a maximum of 3 mg nabilone), with the average dose achieved for nabilone at about 2 mg/day. Seven weeks of treatment were followed by a two-week washout period and seven weeks of alternative treatment. Nabilone not only significantly reduced clinical signs of PTSD (as measured on the Clinical Global Impression scale) and improved well-being (as measured by a well-being questionnaire), but also improved sleep and reduced the number of nightmares. No dropouts from the study and no severe adverse effects were reported. The findings are promising, although the sample was very small. These findings support use of cannabis by PTSD patients to treat their symptoms. However, it is not known whether long-term use of cannabis or THC improves symptoms of PTSD in such patients or worsens it. There is a lack of large-scale, controlled clinical trials that would make a firm conclusion regarding the efficacy or safety of cannabis for the treatment of PTSD possible (Yarnell 2015).

Of most relevance to the treatment of PTSD are the findings that CBD enhanced extinction of fear memories in healthy volunteers (Das et al. 2013). When participants inhaled CBD before or after extinction training in a contextual fear conditioning paradigm, they showed a reduced skin conductance response during reinstatement-enhanced extinction. However, further evidence is required to determine the effect of chronic exposure to CBD on anxiety and /or PTSD in humans. In short, human experimental findings support pre-clinical evidence that CBD has anxiolytic effects with minimal sedative effects and has an excellent safety profile. However, there have been no human clinical trials with CBD as a treatment for PTSD.

Cannabis, Mood, and Depression

Cannabis use has commonly been reported not only to reduce anxiety but also to enhance mood and cause euphoria. Indeed, pre-clinical animal data suggest that THC (at lower doses) and CBD both produce antidepressant effects. This suggests that cannabis use might be effective in reducing depression; however, the reports of cannabis as an antidepressant are contradictory (Patel et al. 2014). Self-report questionnaires examining reasons for cannabis use found that 22 percent of their sample used cannabis to reduce depression (Ware, Adams, and Guy 2005). Indeed, a questionnaire on depressive symptoms in a survey of nearly 4,500 people revealed fewer depressive symptoms in cannabis users than in non-users (Denson and Earleywine 2006). Case reports of five people suffering from depression revealed that depression preceded cannabis use and that the effects of cannabis had some anti-depressant effects (Gruber, Pope, and Brown 1996). In addition, cannabis use is associated with elevated mood and decreased depression in patients with chronic diseases (Patel et al. 2014). On the other hand, oral THC administration to depressed individuals can also result in dysphoria in some patients, especially those who are naive to the psychoactive effects of cannabis. Furthermore, pure THC has been reported to increase anxiety when given alone, whereas co-administration of CBD can counter its effect (Zuardi et al. 1982). As was discussed in chapter 1 without the benefit of the additional cannabinoid compounds (especially CBD) pure THC often does not have the same effect as cannabis consumption.

Endocannabinoids and Anxiety

Long-distance runners describe a "runner's high"—a feeling of euphoria and anxiolysis. It has been generally accepted that endogenous endorphins produce these effects; however, both endorphins and AEA levels are elevated in the blood following running. It is now understood that the anxiolysis after running depends on intact CB_1 receptors in forebrain GABAergic neurons (Fuss et al. 2015).

Surprisingly, since the discovery of the endocannabinoids AEA and 2-AG by Mechoulam's group in the 1990s (Devane et al. 1992; Mechoulam et al. 1995; Sugiura et al. 1995) no study has investigated the effect of systemic administration of these compounds or of inhibitors of the enzymes (FAAH or MAGL) that degrade AEA or 2-AG on anxiety in humans (Mechoulam and Parker 2013). In a pre-clinical animal study, a longer-lasting stable analogue of anandamide (methanadamide) delivered to the prefrontal cortex of rats reduced anxiety-like responses (Rubino and Parolaro 2008).

Because AEA and 2-AG degrade rapidly once released, blocking their enzymatic metabolism (by inhibiting FAAH or MAGL) elevates their endogenous levels for several hours. In contrast with the classical neurotransmitters (as has already been noted), AEA and 2-AG are not stored in synaptic vesicles but rather are synthesized and released in the synaptic cleft after neuronal activation. Presumably, their levels in anxiety and depression, and those of FAAH or MAGL, will be highest in the brain areas involved in the regulation of mood and emotions. Therefore, inhibition of AEA or 2-AG metabolism should enhance CB_1 activation mainly where AEA or 2-AG levels are highest. Such an approach produces elevated levels of endocannabinoids in these regions for as long as 24 hours. Several researchers are currently investigating the effects of FAAH and/or MAGL inhibitors on pre-clinical animal models of anxiety.

Pre-clinical research with animals indicates that inhibition of FAAH, which elevates AEA, reduces anxiety-like and depression-like behavior in animal models as well as reducing isolation-induced vocalizations (Kathuria et al. 2003). This reduction in anxiety-like behavior by FAAH inhibitors occurs only under stressful conditions (Haller et al. 2009). It is also clear that CB_1 antagonists promote anxiety in animal models— probably by blocking the action of AEA, because the anxiety-like behavior is produced only in stressful conditions (Haller et al. 2009; Patel and Hillard 2006). Mice genetically engineered to lack FAAH do not show increased anxiety-like behavior in times of chronic stress, whereas wild-type mice do (Hill, Kumar, et al. 2013). Therefore, FAAH inhibitors, which elevate AEA specifically, clearly have the potential to reduce anxiety in these animal models.

Recent work in Sachel Patel's laboratory (Bluett et al. 2014) has demonstrated an inverse relationship between whole-brain levels of AEA and anxiety-like behaviors in mice exposed to a foot-shock stressor 24 hours earlier. Anxiety levels were enhanced and whole-brain AEA levels were suppressed 24 hours after exposure to foot shock. Pre-treatment with an FAAH inhibitor prevented both the drop in AEA produced by acute stress and the enhancement of the anxiety-like behavior; these effects were mediated by the action of AEA at the CB_1 receptor. A polymorphism in the gene for FAAH that leads to a destabilized FAAH enzyme resulting in increased AEA signaling has been identified (Dincheva et al. 2015). Both humans and mice homozygous in this allele ($FAAH^{A/A}$) show decreased anxiety-like behavior and increased fear-extinction learning. In a human population with this genetic variation, a functional magnetic-resonance imaging (fMRI) investigation revealed increased functional connectivity between the ventromedial prefrontal cortex and the amygdala (brain regions that are involved in emotional behavior) in response to threat, increased fear extinction learning, and decreased anxiety (Dincheva et al. 2015). Together these findings suggest that agents that augment the AEA may be useful in the treatment of stress-related neuropsychiatric disorders. However, there have been no clinical trials with FAAH inhibitors in the treatment of anxiety disorders. There is clearly a need for such studies.

Inhibition of MAGL, which elevates 2-AG, also reduces anxiety in animal models (Kinsey, O'Neal, Long, Cravatt, and Lichtman 2011;

Sciolino, Zhou, and Hohmann 2011; Sumislawski, Ramikie, and Patel 2011). With regard to 2-AG signaling, it has also been reported that deficiency of the 2-AG synthesizing enzyme diacylglycerol lipase-α (DAGL) leads to markedly decreased 2-AG brain levels and to an increase in anxiety-like behavior. The increase in anxiety is rescued by MAGL inhibitors. However, MAGL inhibitors that elevate 2-AG (which, unlike AEA, acts as a full agonist at the CB_1 receptor) have also been shown to produce receptor desensitization (Lutz et al. 2015). Since cannabinoid systems are prominently involved in the modulation of anxiety-related behaviors, these FAAH inhibitors and possibly MAGL inhibitors may provide a novel and beneficial treatment for anxiety.

Stress Coping and the Endocannabinoid System

Stress can be defined as a reaction of the body to a challenge in preparation for possible dangers or injuries. Both physical and psychological stress induce a pattern of responses that allow for coping with the immediate threat. These responses are then followed by recovery to homeostasis. The neural responses to stress are immediate, occurring within seconds. But the hormonal responses mediated by the activation of the hypothalamus-pituitary-adrenal (HPA) axis, with the ultimate release of adrenal glucocorticoids, occur minutes to hours after the stressor. Endocannabinoid signaling in stress-related brain regions is altered by stress.

HPA-axis stress response
The HPA axis provides the body's major neuroendocrine response to stress exposure. (See figure 3.1.) HPA activity promotes a number of processes that meet the energy demands to deal with imminent or perceived threat. In the short term, the HPA-axis stress response is adaptive; however, chronic exposure to stress produces prolonged exposure to stress hormones associated with a number of detrimental consequences, ranging from cardiac diseases to depression (Chrousos 2009).

The typical neuroendocrine HPA-axis response to acute stress that originates in the brain involves activation of corticotropin-releasing hormone (CRH) neurosecretory cells in the paraventricular nucleus (PVN) of the hypothalamus. The resultant release of CRH stimulates the adjacent anterior pituitary gland to secrete adrenocorticotropin-releasing hormone (ACTH) into the circulatory system of the body as a whole. The adrenal glands are endocrine glands located just above the kidneys. The adrenal cortex detects circulating ACTH, which stimulates the

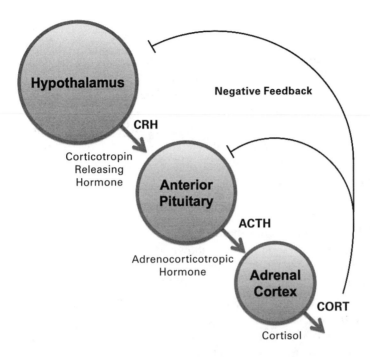

Figure 3.1
The hypothalamic-pituitary-adrenal (HPA) axis. The typical neuroendocrine HPA-axis response to acute stress that originates in the brain involves activation of corticotropin-releasing hormone (CRH) neurosecretory cells in the paraventricular nucleus (PVN) of the hypothalamus. CRH stimulates the adjacent anterior pituitary gland to secrete adrenocorticotropin-releasing hormone (ACTH) into the circulatory system of the body as a whole. The adrenal glands are endocrine glands located just above the kidneys. The adrenal cortex detects circulating ACTH, which stimulates the release of glucocorticoid stress hormones (cortisol in humans, corticosterone in rodents). Cessation of this neuroendocrine stress response occurs when glucocorticoids cross the blood-brain barrier and terminate the further release of CRH and ACTH. Source: Brian M. Sweis, CC BY-SA 3.0, https:// commons.wikimedia.org/w/index.php?curid=23363130

release of glucocorticoid stress hormones—cortisol in humans, corticosterone in rodents. This stress response is regulated by negative feedback from the brain. Cessation of this neuroendocrine stress response occurs when glucocorticoids cross the blood-brain barrier and bind with glucocorticoid receptors found in the PVN of the hypothalamus, in the amygdala, in the prefrontal cortex (PFC), and in the hippocampus. These regions, called the *corticolimbic* regions, regulate emotional behavior. Therefore, the brain controls the termination of the stress response by promoting feedback inhibition of the HPA axis and restoring homeostasis. Although short-term activation of the stress response can be adaptive and beneficial, chronically elevated levels of glucocorticoid hormones from either impaired negative feedback or chronic exposure to stress can lead to many long-term harmful consequences (Chrousos 2009; McEwen 2007). Anxiety and mood disorders are associated with a dysfunctional biological stress response by the HPA axis. Cortisol is elevated in two thirds of depressed individuals. Furthermore, the inability of patients to suppress cortisol after a challenge dose of the steroid dexamethasone is used as a diagnostic tool to determine the presence of depression. Anti-depressant therapies attenuate this hyperactive HPA-axis activity.

Endocannabinoid regulation of the HPA axis

The endocannabinoid system has been shown to regulate the HPA axis in the maintenance of both basal and stress-induced responses (Hill, Patel, et al. 2010). Endocannabinoid signaling inhibits HPA-axis activity, contributing to maintenance of low glucocorticoid levels during basal conditions and restricting HPA-axis activity in acute stress. Mice that have been genetically modified to be deficient in CB_1 receptors have increased basal levels of glucocorticoids and elevated corticosterone secretion. In addition, CB_1 antagonists elevate corticosterone levels in rats. Acute stress reduces the level of AEA in the amygdala and PFC (areas with regulate emotionality) and these changes are accompanied by an elevation of AEA's degrading enzyme, FAAH, mediated by the action of CRH. These reduced AEA levels in the amygdala enable the activation of HPA axis. Treatments that inhibit FAAH (thereby increasing AEA levels) in that region reduce the glucocorticoid response to stress. Indeed, chronic stress results in increase in FAAH and a concomitant reduction in AEA in the amygdala, but these effects are absent in mice genetically engineered to be deficient in FAAH (Hill, Kumar, et al. 2013) Current evidence indicates that AEA contributes to a "tonic like" mechanism, whereas 2-AG promotes a burst-like mechanism in CB_1-receptor activation (Hill and Tasker 2012). That is, AEA is a 'gatekeeper' maintaining basal levels of stress hormones (corticosterone in rodents and cortisol in humans) in the organism, but removal of this AEA-induced tone facilitates activation of the HPA axis and increases emotional and anxiety-like behavior. In essence, the endocannabinoid system facilitates activation of "resilience factors" during or after stressor exposure.

In contrast to AEA, the majority of studies show that acute stress acts to increase 2-AG levels in brain regions relevant to stress. The elevation of 2-AG, however, shows a delay after the stressor. The time lag is consistent with observations that corticosterone levels (which take several minutes to elevate after stressors) mediate the increase in 2-AG after acute stress, by some yet unknown mechanism (Morena, Patel, Bains, and Hill 2016). In addition to acute stress, chronic stress produces enhanced levels of 2-AG and reduced levels of MAGL in the amygdala, which suggested that reduced hydrolysis may account for increased 2-AG (Sumislawski et al. 2011).

Therefore, taken together, the research suggests a bidirectional effect of stress on the endocannabinoid system, with stress exposure first reducing AEA and then increasing 2-AG throughout most brain regions examined. The primary importance of the increase in 2-AG signaling in response to stress seems to be to buffer and constrain the effects of stress on the brain and facilitate termination of the stress response (Morena et al. 2016).

Cannabis and HPA Activity

In contrast to the mood elevating properties of cannabis, studies consistently report that acute cannabis or THC elevates cortisol release in cannabis naive individuals; this effect is blunted in chronic cannabis users. However, chronic cannabis users exhibit elevated basal cortisol levels, which suggests dysregulation of the HPA-axis activity. (See Patel et al. 2014.) Pre-clinical animal studies support the reports in human studies. THC given to rats produces elevated levels of circulating corticosterone (the rat equivalent of cortisol in humans).

The developmental period of adolescence is one of heightened prevalence of emotional and anxiety-related disorders. More than 75 percent of adults with disorders related to fear or anxiety met diagnostic criteria as children or adolescents (Kim-Cohen et al. 2003). Childhood and adolescence also are particularly important in brain development of regions involved in emotional behavior, including maturation of the HPA axis (Crews, He, and Hodge 2007; Guerry and Hastings 2011). The highly plastic adolescent brain may be especially vulnerable to stressors producing impairments to the HPA-axis stress reactivity system that can affect emotionality in adulthood. Pre-clinical animal studies suggest a relationship between high doses of THC during adolescence and the development of depression-like behavior in adulthood. In rats, administration of escalating doses of THC during adolescence results in increased rates of depressive-like symptoms in adulthood (Rubino and Parolaro 2008).

Escalating doses of THC result in downregulation of CB_1 receptors (and thereby reduced endocannabinoid functioning) throughout the limbic regions of the brain involved in emotionality. Indeed, in rats, administration of drugs that elevate endocannabinoid activation of CB_1 receptors can reverse the depressive-like effect induced by adolescent exposure to THC (Realini et al. 2011). However, caution must be taken in applying these findings to humans, insofar as the animals were treated with high doses of pure THC; human cannabis users may be exposed to high concentrations of THC, but they are also exposed to other cannabinoids (e.g., CBD), which may interfere with the effects of THC on the brain. In addition, more recent pre-clinical evidence has revealed that exposure to the CB_1 agonist WIN 55,212,2 during exposure to stressors in adolescence may, surprisingly, reverse the detrimental effects of early-life stress on neurocognitive performance and on enhanced anxiety in adulthood (Alteba et al. 2016).

A number of human studies have suggested a link between heavy cannabis use during adolescence and the development of depression later in life (Bovasso 2001; Patton et al. 2002; Rey, Sawyer, Raphael, Patton, and Lynskey 2002). However, a more recent large (more than 12,000 participants) longitudinal study did not find past cannabis use to be a significant predictor of depression in adults when baseline differences between users and non-users were carefully controlled (Harder, Morral, and Arkes 2006). For instance, individuals who have experienced adversity early in life exhibit an increased risk of developing depression in adulthood and an increased propensity to use cannabis (Hayatbakhsh, Williams, Bor, and Najman 2013). Harder, Morral, and Arkes (2006) suggested, on the basis of their longitudinal results, that a shared genetic predisposition could underlie the association of depression and cannabis use. In support of this hypothesis, among pairs of twins sharing identical genetic and similar environmental conditions, the cannabis-dependent twin exhibits a risk of suicidal ideation or attempts 2.5–3 times that of the non-dependent twin; this risk is even greater if cannabis use began before age 17 (Lynskey et al. 2004). Therefore, genetic factors for cannabis dependence and depression may be moderately correlated. A large meta-analysis of longitudinal and population based studies on cannabis use and risk of affective mental health outcomes concluded that the evidence that cannabis causes depression in humans is not convincing (Moore et al. 2007).

Endocannabinoid Modulation of Extinction of Aversive Memories in Post-Traumatic Stress Disorder

Considerable recent evidence suggests that manipulations of the endocannabinoid system have therapeutic utility in promoting extinction of aversive memories such as those that occur in post-traumatic stress disorder (PTSD). Avoidance of aversive stimuli is crucial for survival of all animals and is highly resistant to extinction. Yet at times aversive memories become haunting, as in PTSD. The endogenous cannabinoid system is specifically involved in extinction learning of aversively motivated learned behaviors. (For a review see Lutz et al. 2015.)

In pre-clinical animal studies of extinction of conditioned fear behavior, classical conditioning techniques are often used. A neutral stimulus (a conditioned stimulus, usually visual or acoustic) is paired with a fear-inducing unconditioned stimulus (e.g., foot shock). After one or more pairings of the conditioned stimulus (CS) with the unconditioned stimulus (US), the presentation of the CS alone evokes a fear response. The association of the two stimuli is consolidated within 6–8 hours of the learning experience and thereby forms a long-term memory. The most thoroughly studied fear response is freezing in rodents. Re-exposure to the CS elicits freezing as the expression of the learning, called the conditioned response (CR). However, re-exposure to the CS in the absence of the US also elicits two opposing neuronal processes: re-consolidation and extinction. Thus, short re-exposure to the CS triggers another round of memory consolidation (re-consolidation) in which new information can be integrated into the original memory. On the other hand, prolonged or repeated exposure to the CS in the absence of the US triggers extinction, resulting in a reduction in the expression of fear. Extinction is believed to be impaired in patients suffering from specific fear-related disorders such as phobias or PTSD. Therefore, any manipulations that can enhance extinction will have profound clinical relevance.

Marsicano et al. (2002) reported that CB_1 knockout mice and wild-type mice administered the CB_1 inverse agonist/antagonist rimonabant showed impaired extinction in classical auditory fear-conditioning tests, with unaffected memory acquisition and consolidation. That effect was mediated by blockade of elevated AEA in the basolateral amygdala (BLA) during extinction. Indeed, recent evidence (Kamprath et al. 2011) indicates that infusion of CB_1 antagonists into the BLA impairs long-term extinction memory. In addition, Varvel and Lichtman (2002) reported that CB_1 knockout mice and wild-type mice exhibited identical acquisition rates in learning to swim to a fixed platform in a Morris water maze; however, the CB_1-deficient mice demonstrated impaired extinction of the originally learned task when the location of the hidden platform was moved to the opposite side of the tank. Because animals deficient in CB_1-receptor activity show impairments in suppressing previously learned behaviors, CB_1 agonists would be expected to facilitate extinction of learned behaviors in non-deficient animals—something that has been reported repeatedly (Mechoulam and Parker 2013).

Since CB_1 agonists facilitate extinction of aversive memories, it might be expected that treatments that enhance endogenous levels of AEA and/or 2-AG would also be potential treatments for PTSD. Chhatwal et al. (2005) reported that the re-uptake blocker (and FAAH inhibitor) AM404 selectively facilitated extinction of fear-potentiated startle in rats—an effect that was reversed by rimonabant pre-treatment (Chhatwal, Davis, Maguschak, and Ressler 2005). These effects appear to be specific to extinction of aversively motivated behavior, because neither CB_1-deficient mice (Holter et al. 2005) nor wild-type mice treated with rimonabant (Niyuhire, Varvel, Thorpe, et al. 2007) displayed a deficit in extinction of operant responding reinforced with food. FAAH inhibition also promoted extinction of a conditioned place aversion produced by naloxone-precipitated morphine withdrawal, but did not promote extinction of a morphine-induced or amphetamine-induced conditioned place preference (Manwell et al. 2009). Thus, elevation of AEA selectively facilitates extinction of aversive memories. These effects may be mediated by enhanced AEA signaling in the BLA (Lutz et al. 2015).

The literature regarding the effect that 2-AG may have on fear extinction is controversial. Recently it was shown that mice deficient in DAGLα, the precursor to 2-AG, showed no impairments in fear acquisition, but showed impaired fear extinction (Jenniches et al. 2016), suggesting that 2-AG is required for fear extinction. On the other hand, treatment of mice with the MAGL inhibitor JZL184 (which elevates 2-AG) was also shown to promote fear expression and to impair fear extinction, by the action of 2-AG on CB_1 receptors in forebrain GABAergic neurons (Llorente-Berzal et al. 2015). Future research may reconcile these contradictory effects of 2-AG and extinction of conditioned fear; however, one possibility is that increased AEA levels mediate acute fear relief but increased 2-AG levels promote the expression of conditioned fear by blocking GABA release. In view of the dual effects of 2-AG on glutamate and GABA, it is possible that optimal 2-AG signaling is required for processing of fear responses (Lutz et al. 2015).

It has been well established that extinction is not unlearning but instead is new inhibitory learning that interferes with the originally learned response (Bouton 2002). The new learning responsible for extinction of aversive learning appears to be facilitated by AEA and to be prevented by inhibition of the endocannabinoid system. Some suggest that the apparent effects of manipulation of the endocannabinoids on extinction may actually reflect their effects on re-consolidation of the memory that requires its re-activation (Lin et al. 2006; Suzuki, Mukawa, Tsukagoshi, Frankland, and Kida 2008). Whenever a consolidated memory is recalled, it switches to a labile state and undergoes re-consolidation if its processing is not disrupted. Depending on the conditions of retrieval and the strength of the original trace, these re-activated memories can undergo two opposing processes: re-consolidation (when the conditions favor the permanence of the trace) and extinction (when the conditions indicate that the memory has no reason to persist). Suzuki et al. (2008) have proposed that the endocannabinoid system is important for the destabilization of re-activated fear memories; that is, that fear memory cannot be altered during re-stabilization if it was not previously destabilized via activation of the CB_1 receptor. Whatever the actual mechanism for facilitated extinction with activation of the endocannabinoid system and inhibited extinction with inhibition of the endocannabinoid system, these results have considerable implications for the treatment of PTSD. Progress in the development of drugs that enhance AEA signaling will be of great benefit in the treatment of this distressing disorder. However, there have been no human clinical trials of treatment of PTSD with FAAH inhibitors.

The promise of cannabinoid treatment of PTSD was recently tested experimentally in humans (Rabinak et al. 2014). The participants were conditioned to fear two conditioned visual stimuli (CS+) on a computer screen that were paired with an aversive burst of white noise. Another visual stimulus (CS-) was not paired with the white noise. On the next day, the participants underwent an extinction session during which one of the CS+ stimuli was extinguished and the other CS+ was not presented. Two hours before that extinction session, participants ingested a 7.5-mg/kg THC capsule or a placebo capsule. To assess extinction retention, a test of memory recall was administered 24 hours after the extinction session (day 3). The participants receiving the THC capsule had better retention of the extinction memory, although they did not differ in recall of the non-extinguished CS+. This suggests that THC may selectively enhance memory of fear extinction in humans, as was suggested by the pre-clinical animal work. A similar effect was been reported with CBD given to healthy volunteers after extinction learning—that is, CBD also enhanced extinction consolidation (Das et al. 2013). Rabinak et al. (2014) reported that THC enhanced the activity of the ventromedial prefrontal cortex and the hippocampus (two regions involved in fear extinction in humans) during extinction of fear conditioning in an fMRI study, using a randomized, double-blind, placebo-controlled design to test their hypothesis. This is the first evidence that THC modulates prefrontal-limbic circuits during fear extinction in humans. It suggests that cannabinoid agonists may correct the impaired behavioral and neural function during extinction recall in patients with PTSD and other fear-learning-related disorders (Rabinak et al. 2014). Treatments that selectively facilitate fear extinction may have great benefit in aiding patients with PTSD to live more normal lives devoid of the distressing emotional memories of trauma.

Cannabis and Sleep Disorders

Intoxication from cannabis is most commonly described as a feeling of relaxation, and many users report smoking marijuana to help them sleep; however, few studies have focused specifically on the relationship between cannabis use and sleep (Gates, Albertella, and Copeland 2014). Disorders of sleep are generally treated with sedatives such as benzodiazepines-receptor agonists and non-benzodiazepine hypnotic agents. However, side effects, including dependence and tolerance, occur with those drugs (Murillo-Rodriguez, Aguilar-Turton, Mijangos-Moreno, Sarro-Rimirez, and Arias-Carrion 2014).

Some evidence suggests that the endocannabinoid system plays a role in sleep. Participants deprived of sleep for 24 hours had increased levels of OEA (a natural analogue of AEA) in their cerebrospinal fluid, but not in serum, whereas levels of AEA were not changed (Koethe, Schreiber, et al. 2009). In rats, both acute and sub-chronic administration of AEA induce sleep (Herrera-Solis, Vasquez, and Prospero-Garcia 2010).

Early investigations of cannabis use and sleep were conducted in the 1970s. It has been demonstrated that daily doses of 70–210 mg of THC induce sleep in humans (Feinberg, Jones, Walker, Cavness, and Floyd 1976; Feinberg, Jones, Walker, Cavness, and March 1975). As for the effects on sleep architecture, the most consistent finding from these early studies using polysomnograph technology was a reduction in rapid-eye-movement (REM) sleep, the stage in which dreaming occurs. However, the interpretation of these findings was limited because the samples were small (Schierenbeck, Riemann, Berger, and Hornyak 2008). More recently, human clinical trials in which sleep was a secondary measure of positive treatment outcomes of medical cannabis have contributed to our understanding of the effects of cannabis on sleep. Several such trials found that nabiximols (Sativex 2.7 mg THC: 2.5 mg CBD/spray) to reduce spasticity and pain was likely to improve subjective sleep parameters but was unlikely to change sleep stage architecture.

The current understanding of cannabis use and sleep is somewhat inconsistent and often lacks statistical control for confounding factors. For instance, in one set of findings medicinal users report that cannabis use alleviates sleep problems, whereas cannabis use is also reported as a risk factor for sleep disorders (Gates et al. 2014). A recent systematic review of the literature indicated that a total of eleven studies (with of 203 participants in all) investigated the effect of recreational cannabis use on sleep. The risk of bias in these studies was high, most commonly because of confounding factors such as pre-existing sleep problems, gender, and age. Cannabis dose (and strain) and dosing duration varied in these studies. Six of these studies used objective EEG measures. Only one of these studies was conducted in the past decade, and the results were inconsistent. A study by Nicholson, Turner, Stone, and Robson (2004) was rated as having the highest quality. Four male and four female occasional cannabis users participated in a double-blind, placebo controlled cross-over trial. The cannabinoids were delivered by sublingual spray. In a within-subject design, each participant received a single dose of 15 mg THC, 5 mg THC+5 mg CBD, 15 mg THC +15 mg CBD, and placebo 22½ hours before the lights were turned out. The participants abstained from alcohol and caffeine. Two of them continued to smoke tobacco. Nicholson et al. found that THC was sedating but CBD was alerting. This finding is consistent with the pre-clinical animal literature, with CBD as a wake-promoting factor. Intracerebroventricular administration of CBD (10 micrograms) to rats at the beginning of the lights-on period increased wakefulness and decreased REM sleep (Murillo-Rodriguez, Millan-Aldaco, Palomero-Rivero, Mechoulam, and Drucker-Colin 2006).

A total of 28 studies included a measure of sleep as a treatment outcome for various illnesses assessed, including pain, multiple sclerosis, anorexia, cancer, and HIV (Gates et al. 2014). These studies included assessments of Marinol, nabilone, and Sativex. Although the majority of the studies did not include validated measures of sleep, 22 of them reported a significant and positive effect on sleep in the clinical trial. However, it is not clear whether this was a result of the mediating improvement of the underlying condition being treated (e.g., pain) that interfered with sleep as a pre-existing condition. Patients diagnosed with PTSD displayed significant improvements in sleep time and quality of sleep when treated with nabilone (Fraser 2009). Patients with fibromyalgia (a disease characterized by chronic pain and insomnia) who were treated with nabilone (0.5–1 mg before bedtime) showed significant improvement in sleep (Ware, Fitzcharles, Joseph, and Shir 2010). However, the mechanism by which CB_1 agonists improve sleep is unknown. As Gates et al. (2014) suggest, there is clearly a need for large-scale, longitudinal, well-controlled studies of the specific effects of cannabinoids on sleep.

Conclusion

The psychoactive cannabinoid THC exerts a biphasic effect on anxiety and mood. Low doses reduce anxiety and elevate mood, but chronic exposure to high doses increase anxiety and can result in depressive symptomology. Treatment of healthy individuals with a CB_1-receptor antagonist increases indices of anxiety, depression, and suicidal ideation (Nissen et al. 2008; Hill and Gorzalka 2009). Individuals with major depression or PTSD after the terrorist attacks of September 11, 2001 have reduced levels of endocannabinoids in their circulating blood supply, and there is a high negative correlation between endocannabinoid levels and anxiety measures in humans (Hill, Bierer, et al. 2013; Hill, Miller, Ho, Gorzalka, and Hillard 2008). Thus, a compromised endocannabinoid system may be unable to restore homeostatic balance, and the result may be symptoms of anxiety or depression.

Extinction learning, most relevant for preventing recall of memories of traumatic events, is particularly sensitive to actions of CB_1 agonists and FAAH inhibitors. The rich and consistent pre-clinical literature demonstrates that those agents strengthen extinction learning for aversive events, which interferes with subsequent recall of the originally learned memory. Human research is beginning to yield results consistent with animal research. Therefore, the evidence clearly points to the use of cannabinoid treatments to promote the extinction of traumatic memories in human patients with PTSD. The potential of THC to facilitate sleep and reduce nightmares in PTSD patients is also promising (Fraser 2009). Considerable pre-clinical evidence indicates that treatments which elevate AEA by inhibiting its degradation by FAAH may be highly effective in reducing anxiety and in promoting extinction of aversive memories in PTSD. However, there has not been a single published report of a human clinical trial with these promising compounds for these disorders.

The pre-clinical data argue strongly for human clinical trials of the use of FAAH inhibitors for anxiety/depression. The recent human work showing the CBD also facilitates extinction learning in humans provides support for the use of nabiximols (Sativex) as well, although there is as yet no pre-clinical evidence of combined THC and CBD facilitation of extinction learning.

4

Cannabinoids and Psychosis

Probably the greatest health concern about the use of cannabis is the putative link to later development of psychotic disorders. Several population-based studies and a few longitudinal studies show that regular cannabis use in adolescence is associated with about a twofold increase in the risk of psychosis (Moore et al. 2007), and regular cannabis use in adolescence has been linked to an earlier age of onset of psychosis (Large et al. 2007). However, there is considerable controversy about whether these associations are causal. Clearly, the overwhelming majority of people who have used cannabis do not develop a psychotic disorder, and most people who develop a psychotic disorder may never have used cannabis. Cannabis use is neither necessary nor sufficient to "cause" schizophrenia. However, it may contribute to the risk of the development of a psychotic disorder in vulnerable individuals.

Schizophrenia is a poorly understood condition with no clearly known modifiable risk factors. It has a worldwide incidence of approximately 1 percent, and its onset occurs most commonly in late adolescence or early adulthood. It is clear that THC can produce transient, acute, but usually mild psychotic experiences in an intoxicated user (D'Souza et al. 2004) and can worsen the symptoms in people with schizophrenia (Di Forti, Morrison, Butt, and Murray 2007). On the other hand, a recent a meta-analysis showed that people diagnosed with schizophrenia who use cannabis function better cognitively than individuals with schizophrenia who do not use cannabis (Yucel et al. 2012). The big question is whether cannabis use (especially in adolescence) increases the risk of later development of schizophrenia and other psychotic disorders. If it does, then cannabis use may be one of the few modifiable risk factors for schizophrenia (Gage, Munafo, MacLeod, Hickman, and Smith 2015).

Acute Psychosis Associated with Cannabis Intoxication

It has long been known that, cannabis intoxication (more specifically, THC intoxication) can elicit paranoid psychosis in some users (Murray, Morrison, Henquet, and Di Forti 2007). This effect of cannabis has been known for thousands of years, and has been described in relatively recent reports of laboratory experiments (D'Souza et al. 2005). Jacques-Joseph Moreau reported in 1845 that hashish could induce "acute psychotic reactions, generally lasting but a few hours, but occasionally as long as a week; the reaction seemed dose-related and its main features included paranoid ideation, illusions, hallucinations, delusions, depersonalization, confusion, restlessness and excitement" (Radhakrishnan, Wilkinson, and D'Souza 2014). Indeed, a number of reports suggest that heavy cannabis use can produce psychotic reactions in healthy people and may worsen a pre-existing chronic psychotic disorder (Di Forti et al. 2007).

Experimental studies using a randomized-control paradigm with human participants provide opportunities to control a number of variables that are not controlled in anecdotal case reports. D'Souza et al. (2004) characterized the positive psychoactive symptoms, negative symptoms, and cognitive effects of intravenous THC in healthy individuals. In a double-blind randomized placebo-controlled study, they administered THC intravenously at 2.5 and 5 mg/kg to 22 healthy adults. Participants were screened to rule out psychiatric or family history of psychotic disorders. When assessed by the positive and negative syndrome scale for schizophrenics (Kay, Opler, and Lindenmayer 1988), THC produced transient (up to 60 minutes) positive psychotic symptoms (hallucinations, delusions, formal thought disorder, bizarre disorganized behavior) and negative symptoms (e.g., affective flattening, an attentional impairment involving volition-apathy and anhedonia) as well as perceptual alterations. However, since THC also produces known sedative and cataleptic effects in animal studies, it is not clear if some of the negative symptoms are independent of THC-induced sedation. It is interesting to note that the positive psychomimetic effects of THC were not alleviated by anti-psychotic treatment (dopamine D2-receptor antagonists) and other studies showed that chronic anti-psychotic treatment failed to protect schizophrenia patients from the symptom-exacerbating effects of THC (D'Souza 2007). More recently, an fMRI study (Fusar-Poli et al. 2009) found that, whereas ingestion of 10 mg THC resulted in increased skin conductance responses during processing of fearful faces, ingestion of 600 mg CBD led to a reduction in anxiety and a decrease in skin conductance response. In addition, THC and CBD have been found to have opposite effects on blood-oxygen-level-dependent (BOLD) responses in tasks of verbal recall, in response inhibition, and in the processing of fearful expressions (Bhattacharyya et al. 2010). Therefore, CBD may actually reduce the acute psychotic effect of THC.

The likelihood of experiencing the psychotic effects of acutely administered THC is also modulated by personality and genetic factors. Catechol-O-methyl transferase (COMT) is an enzyme that metabolizes dopamine in the prefrontal cortex. Individuals with the Val/Val genotype for COMT have a higher metabolic activity of that enzyme (and thus a lower level of prefrontal cortical dopamine) than individuals with the Met/Met polymorphism. Cécile Henquet and colleagues evaluated the effects of the interaction of COMT polymorphism and a trait index of psychosis liability of smoked THC (0.3 mg/kg) on cognitive performance and psychosis in 30 healthy individuals. (Henquet, Di Forti, Morrison, Kuepper, and Murray 2008; Henquet et al. 2006). Individuals with the Val/Val polymorphism and high scores on psychosis liability had higher THC-induced psychotic symptoms than individuals with the Met/Met polymorphism.

THC has also been shown to induce abnormalities in electrophysiological indices of brain function known to be present in schizophrenia. In animal studies THC has been shown to disrupt synchronized neural oscillations in the hippocampus, an effect that can be reversed by CBD (Wilkinson, Radhakrishnan, and D'Souza 2014). The pattern of brain activity in schizophrenic patients shows elevated random neural activity, which has also been produced by THC in a dose-dependent manner in normal healthy participants (Skosnik, Cortes-Briones, and Hajos 2016). These increases in neural noise were related to the positive symptoms, not the negative symptoms of psychosis. Drugs that induce psychosis may, therefore, do so by increasing the level of task-irrelevant random neural activity (neural noise).

Schizophrenia involves dysregulation of dopamine transmission. Evidence regarding the effects of THC on central dopamine transmission has been equivocal. In their report of the first human study using positron-emission tomography (PET), Bossong et al. (2009) noted that THC produced a slight increase in striatal dopamine. However, others failed to replicate that finding (Barkus et al. 2011; Stokes, Mehta, Curran, Breen, and Grasby 2009). Bloomfield et al. (2014) reported a reduction in dopamine release in cannabis users, which is inconsistent with the higher levels of dopamine release observed in people with psychosis.

In summary, THC intoxication can occasionally produce several symptoms of schizophrenia. These effects are dose-related and are specific to THC but not to other cannabinoid constituents. CBD may actually counteract these effects. However, these effects do not last beyond the period of intoxication; indeed, in laboratory studies these effects can be resolved in 2–4 hours (D'Souza et al. 2004).

Cannabis and Persistent Psychotic Disorders

A lingering and most important question is whether cannabis exposure (especially in adolescents) can increase the risk of later developing psychotic disorders, including schizophrenia. Several epidemiological studies and meta-analyses have shown an association between cannabis use in adolescence and development of schizophrenia, the risk increasing with higher frequency of cannabis use (Andreasson, Allebeck, Engstrom, and Rydberg 1987; Arseneault, Cannon, Witton, and Murray 2004; Zammit, Allebeck, Andreasson, Lundberg, and Lewis 2002). One of the first studies demonstrating a link between frequency of cannabis use and psychotic diagnosis was a longitudinal 15-year cohort study of Swedish military conscripts. A dose-response relationship was observed between self-reported cannabis use by age 18 and psychiatric hospitalization for schizophrenia by age 45, with a threefold increase in risk in those who used cannabis more than 50 times by age 18 (Andreasson et al. 1987; Zammit et al. 2002). It is important to note, however, that among the more than 50,000 conscripts in this initially very large sample only 362 (0.72%) were later hospitalized for schizophrenia. The probable reason that the percentage was lower than the typical 1 percent was earlier screening for mental illness by the military. Of the men who later developed schizophrenia, only 18 had used cannabis and no other drugs. Of these 18 cannabis-only users, only four reported having used cannabis more than 50 times before being conscripted (Zammit et al. 2002). One might also predict that young military men may be likely to under-report their use of cannabis.

Since the initial Swedish conscript study, a number of other large, long-term, cross-sectional epidemiological studies have reported a link between cannabis use in adolescence and subsequent hospitalization for schizophrenia. A meta-analysis based on a thorough review of the existing literature and published by Theresa Moore and colleagues (2007) reported a modest but consistent 40 percent increase in the risk of any psychotic outcome in cannabis users compared with nonusers and a 100 percent (or twofold) increase among more frequent cannabis users (Moore et al. 2007). In addition, early cannabis use (before age 15) confers greater risk for schizophrenia outcomes than later cannabis use (by age 18), which may also be a result of more prolonged exposure to cannabis during adolescent years (Arseneault et al. 2004). However, the extent to which cannabis is causative in this association is hotly debated. There is more agreement with the findings that heavy use of cannabis (and other illicit substances) may result in a psychotic episode occurring earlier (by 2.7 years) in young people who are vulnerable to psychosis for various genetic and environmental reasons (Large, Sharma, Compton, Slade, and Nielssen 2011).

The demonstration of an association between cannabis use and schizophrenia does not necessarily mean that cannabis use *causes* schizophrenia. Such observational data are correlational and are subject to alternative interpretations, including reverse causality (people with an existing psychosis are more likely to use cannabis), bias (where problems with measurement or sample selection lead to incorrect estimates), and confounding (where other variables that increase risk of both cannabis use and psychosis lead to spurious associations) (Gage et al. 2015). Most of the longitudinal follow-up studies that have been conducted to date have tried to account for reverse causation by excluding individuals with baseline psychotic symptoms or by adjusting for baseline symptoms (Castle 2013; Fergusson, Horwood, and Ridder 2005; Zammit et al. 2002), thus weakening the reverse causality argument. However, pinpointing the onset of illness and thus establishing the temporal relationship between cannabis exposure and schizophrenia can be difficult (Wilkinson et. al. 2014). In fact, an earlier prospective study of a large population of Dutch teenagers noted a bidirectional association such that vulnerability to psychosis at ages 13 and 16 predicted cannabis use at ages 16 and 19 respectively (Linszen, Dingemans, and Lenior 1994). Residual confounding beyond that accounted for in the studies probably would lead to an overestimate of the causal association, insofar as individuals who use cannabis regularly and those at higher risk of developing mental illness share similar characteristics (Gage et al. 2015). In addition, sample-selection bias (which is difficult to control when cannabis use is determined by self-reporting) is a source of potential measurement error. Furthermore, cannabis comes in a variety of strains that have different ratios of THC to CBD, which may result in measurement error when the ratio in the sample is not known. And in view of the long half-life of THC, heavy users of cannabis may be rarely unintoxicated; that might lead to misclassification of psychotic episodes, which may actually reflect acute effects of THC when participant samples are not limited to former users (Gage et al. 2015).

Cannabis use and risk factors for psychosis

In determining the potential of cannabis use as a risk factor for schizophrenia, it is important to identify additional factors that may interact with cannabis use and promote mental illness. There is evidence of an additive interaction of environmental risk (e.g., maltreatment as a child) and cannabis use in the development of schizophrenia (van Os and Kapur 2009), but no evidence of a synergistic effect. Why do only a minority of cannabis users develop psychosis? One factor could be particular genetic variants that may modify the risk of cannabis' promoting psychosis. Initial investigations implicated an interaction between cannabis and a functional polymorphism in the gene for COMT. Caspi et al. (2005) reported that adolescent cannabis users (but not individuals who began using cannabis as adults) with the Val/Val COMT genotype were 10 times as likely to develop psychosis as those with the Met/Met genotype. This probably was not a case of reverse causality, because there was no increase in risk of psychosis in the Val/Val COMT genotype participant who did not use cannabis, nor were the Val allele carriers more likely to use cannabis. Although initial laboratory and community-based experimental studies supported this link (Henquet et al. 2009; Henquet et al. 2006), more recent epidemiological studies (Zammit, Owen, Evans, Heron, and Lewis 2011; Zammit et al. 2007) have not supported it. Interestingly Costas et al. (2011) showed that schizophrenic patients who were Met homozygotes had higher rates of lifetime cannabis use than Val homozygotes. These data are at odds with the earliest report (Caspi et al. 2005) that the Val/Val genotype may serve as a genetic marker of a proclivity to develop schizophrenia after cannabis use in adolescence.

Recently another potential genetic marker has been identified: polymorphic variation in the gene for the intracellular enzyme AKT1, which codes for a phosphorylating enzyme that has been shown to be activated by CB_1 (Sanchez, Ruiz-Llorente, Sanchez, and Diaz-Laviada 2003). Daily cannabis users who had the cytosine-cytosine (C/C) genotype were twice as likely as those with the thymine-thymine (T/T) genotype to develop a psychotic disorder (Di Forti et al. 2012; van Winkel et al. 2011). Also, individuals with the C/C genotype who used cannabis had a higher risk of psychotic disorder than individuals who did not use cannabis. Recent studies have identified other genes that may moderate the association between cannabis use and psychosis, including the genes for brain derived neurotropic factor (BDNF) (Decoster et al. 2011) and the dopamine transporter DAT1 (Bhattacharyya et al. 2012), which removes synaptic dopamine in striatal regions; and polymorphisms of the gene for the CB_1 receptor CNR1 (Ho, Wassink, Ziebell, and Andreasen 2011). These findings are particularly important: If a genetic marker exists for the potential of cannabis use to trigger a later psychosis, genetic screening can be done to identify individuals at high risk.

The results supporting the hypothesis that some gene variants may interact with cannabis use to promote the development of schizophrenia are tentative. An alternative explanation is that a shared genetic link makes individuals more likely to develop schizophrenia and use cannabis. A recent report from a large genome-wide study of 2,082 healthy individuals suggests that the association between schizophrenia and cannabis use may be due in part to shared genetic etiology across common variants. Power et al. (2014) found that individuals with an increased genetic predisposition to schizophrenia are both more likely to use cannabis and more likely to use it in greater quantities. Therefore, it remains unclear whether cannabis use directly increases the risk of psychosis or whether the same genes that increase psychosis risk may also increase the risk of cannabis use. That is, this gene-environment correlation allows individuals to "choose and shape their own environment based on their own innate preferences" (Power et al. 2014, p. 1202).

Adolescent cannabis use as a risk factor for psychosis

There is growing evidence that a greater risk of psychosis is evident when use begins during adolescence than when it begins in adulthood (Arseneault et al. 2004; McGrath et al. 2010; Schubart, van Gastel, et al. 2011). Indeed, one study reported that the association between cannabis and psychotic disorders was significant only when cannabis use began before age 14 (Schimmelmann et al. 2012). Another indicated that daily cannabis use was associated with more psychotic experiences only among individuals who began using cannabis before the age of 17 (Ruiz-Veguilla et al. 2013). Insofar as some regions of the brain are still developing during adolescence, the effects of cannabis exposure during that period may affect the developing neural networks. Some animal studies (O'Shea, McGregor, and Mallet 2006) suggest that CB_1 agonists produce greater effects on cognition and social interaction if the drug is administered during adolescence than when it is administered during adulthood

(Quinn et al. 2008; Rubino and Parolaro 2008; Schneider and Koch 2003); however, it is difficult to model a complex disorder such as schizophrenia with currently available animal models. Another interpretation of increased risk of psychosis among adolescent cannabis users is that cumulative exposure may be more relevant than age of onset of cannabis use, with earlier cannabis use resulting in greater cumulative exposure. Stefanis et al. (2013) found a consistent lag of 7–8 years between age of onset of cannabis use and age of onset of psychosis in a retrospective study of 997 individuals ages 12–19 years at the onset of cannabis use. Cannabis use is also linked with an earlier age of onset of psychosis— earlier by 2.7 years (Large et al. 2011).

Cannabis strains and psychosis

Different types of cannabis contain different quantities of THC, the psychoactive ingredient that may be linked to risk factors in adolescents. A popular strain of cannabis used recreationally is the pungent "skunk," which is specifically bred to have very high THC content and little or no CBD. The use of "skunk" may present a higher risk for psychosis than cannabis products that were available before there was intensive selective breeding for high THC content. A widely cited study conducted in London found that patients experiencing their first psychotic episode and healthy matched controls were equally likely to report ever having used cannabis. However, the probability of reporting a psychotic disorder was seven times as high in patients who used "skunk" cannabis as in those who used hashish resin that contained both THC and CBD (Di Forti et al. 2009). CBD has been shown to have anti-psychotic effects (Leweke et al. 2012) on its own and has also been shown to counteract some effects of THC. Therefore, it is not simply the higher THC content in "skunk" that is the potential risk factor, because users probably self-titrate their intake (Freeman et al. 2014); rather, it is the lack of CBD in "skunk." The marijuana used before selective breeding for high THC contained a higher ratio of CBD to THC. Indeed, recent epidemiological studies have shown that the lower ratio of CBD to THC in current commercially grown cannabis products is associated with more positive psychotic symptoms (Morgan and Curran 2008; Schubart, Sommer, et al. 2011).

CBD as anti-psychotic

Recent research has identified CBD as a potential anti-psychotic agent. In pre-clinical animal models CBD has been shown to reduce psychosis-like behavior in both dopamine-based and glutamate-based laboratory models of schizophrenia symptoms (Parolaro, Zamberletti, and Rubino 2014), and the prevalence of cannabis-linked psychosis is lower when street cannabis contains a higher proportion of CBD (Di Forti et al. 2009). In healthy humans, CBD reverses THC-induced psychotic symptoms and ketamine-induced depersonalization (a human glutamate model of psychosis). In a controlled clinical trial that compared CBD and the standard anti-psychotic amisulpride in 33 patients over four weeks (Leweke et al. 2012), both groups showed significant improvements from baseline in the positive and negative syndrome scale (PANSS); however, CBD showed a superior safety profile, without extrapyramidal symptoms, weight gain, or elevated serum prolactin. CBD has also been demonstrated to be effective in treatment of psychotic symptoms in patients with Parkinson's Disease. In addition to reducing psychosis, CBD ameliorated motor symptoms (Zuardi et al. 2009). These results suggest that CBD, which repeatedly has been shown to lack toxicity and to produce no adverse side effects in normal healthy participants (in contrast with the usual anti-psychotic drugs), may ameliorate psychotic symptoms in schizophrenic patients. There is an urgent need to explore more fully if and how CBD alleviates both the positive and negative symptoms of schizophrenia (Parolaro et al. 2014).

Endocannabinoids and schizophrenia

There is a link between the level of anandamide (AEA) in the cerebral spinal fluid and symptoms of schizophrenia in human patients. AEA levels in the blood and in the cerebral spinal fluid (Koethe, Giuffrida, et al. 2009) are higher in schizophrenic patients than in healthy volunteers, and remission is associated with a decrease in these levels (De Marchi et al. 2003). In fact, AEA concentrations are higher by a factor of 8 in drug-naive paranoid schizophrenics (Giuffrida et al. 2004; Hillard, Weinlander, and Stuhr 2012). The increase in AEA occurs very early in the course of the disease (the prodromal phase), and individuals with the lowest AEA concentrations are at highest risk of progressing to a psychotic state (Koethe, Giuffrida, et al. 2009). These data led to the hypothesis that elevated AEA concentrations are protective against the symptoms of schizophrenia. It is interesting to note that frequent cannabis users who were schizophrenics were found to have significantly lower AEA concentrations than schizophrenics who used cannabis infrequently (Leweke et al. 2007).

Cannabis and Schizophrenia

The majority of people who use cannabis do not develop schizophrenia. It is clear however, that cannabis can elicit acute psychosis, can worsen the course of pre-existing schizophrenia, and may be a trigger in the development of schizophrenia in at-risk populations. Most of the longitudinal, case-control, and cross-sectional studies conducted to date have found consistent evidence of an association between cannabis use and psychosis, even after adjustment for covariates. There also was evidence of such an association when dose-response relationships were assessed: Acute exposure to high doses of THC produced psychotic experiences that could progress to chronic cannabis disorders. Indeed, some studies found evidence that cannabis use specifically, and not use of other drugs, is associated with psychosis. On the other hand, insofar as cannabis use has increased greatly over the past 70 years in Western countries, an argument against a causal association between cannabis and schizophrenia is that a corresponding increase in schizophrenia diagnoses has not been observed (Frisher, Crome, Martino, and Croft 2009). In the United Kingdom, for instance, cannabis use has risen by a factor of 10–20 since the 1970s (Hickman et al. 2009); however, recent data have shown no increase in new admissions for schizophrenia; instead, there is a suggestion of a decrease (Frisher et al. 2009). A similar pattern has been reported in Australia (Degenhardt, Hall, and Lynskey 2003). Such ecological studies have found little association between increased cannabis use in recent decades and incidence of psychotic disorder (Gage, Zammit, and Hickman 2013).

Overall, the evidence suggests that the association between cannabis exposure and schizophrenia is small but consistent (Wilkinson et al. 2014). Given that the overall incidence of schizophrenia among the general population is 1%, if this association is causal and the magnitude is the approximately 2 fold in regular cannabis users, as is commonly accepted in the literature, then 98 percent of regular cannabis users will not develop schizophrenia (Gage et al. 2015). Very large numbers of cannabis users would have to be prevented from using cannabis to change the incidence of schizophrenia significantly. The absolute number of cannabis users that would have to stop using cannabis to prevent one case of schizophrenia per year has been estimated at approximately 5,000 for heavy users among men and 10,000–15,000 among women (Hickman et al. 2009). However, those at higher genetic risk or those that use more potent strains of high-THC cannabis may be at greater risk. Cannabis use may indeed be harmful, especially to adolescents vulnerable to development of schizophrenia, but it is also important to avoid overstating its harmfulness, as was done in the days of the movie *Reefer Madness*. The message may be ignored when experience fails to match the warnings given to young people (Gage, Hickman, and Zammit 2016). Nonetheless, the bulk of the literature suggests that individuals with a family history of schizophrenia, individuals with prodromal symptoms, and individuals who have experienced discrete episodes of psychosis related to cannabis should be strongly discouraged from using THC-predominant cannabis and psychoactive cannabinoids (Wilkinson et al. 2014).

5

Cannabinoids, Learning, and Memory

Cognition involves the ability to acquire, store, and later retrieve new information. It is clear that THC, the chief psychoactive component in cannabis, produces acute cognitive disturbances in humans and in animals, affecting short-term memory more profoundly than long-term memory. Deficits are seen in the ability to simply hold information for brief periods—for example, to remember a telephone number before dialing it. However, the effect of acute exposure to THC on working memory is short-lived. A controlled study found no residual cognitive impairment 24 hours after ingestion of THC (Curran, Brignell, Fletcher, Middleton, and Henry 2002). Whether chronic exposure to marijuana produces long-term cognitive deficits during abstinence is a matter of controversy in the literature.

Acute Effects of Cannabis on Cognition in Humans

Marijuana produces a transient impairment in short-term memory and in consolidation of short-term memories into long-term memory; however, it does not produce an impairment in retrieval of information once it has been previously encoded into long-term storage. These effects depend on the presence of substantial THC in marijuana. A recent naturalistic study revealed that cannabidiol prevented the memory-impairing effects of acute THC in humans (Morgan, Schafer, Freeman, and Curran 2010). Therefore, a change in the relative ratio of THC to cannabidiol in cannabis will profoundly alter the effects of cannabis on memory in human marijuana smokers.

Cannabis use has been associated with impaired cognition during acute intoxication; however, whether the impairment persists beyond the intoxicated state in long-term users is not as clear. Recently a systematic review of studies of the acute and chronic effects of cannabis and cannabinoids on neuropsychological task-based measures of cognition and on persistence or recovery after abstinence was conducted (Broyd, van Hell, Beale, Yucel, and Solowij 2016). Results from 105 studies that met the authors' criteria for inclusion were reviewed. Broyd et al. (2016) concluded that verbal learning and memory and attention are most consistently impaired by acute and chronic exposure to cannabis (intravenous THC, vaporized cannabis, and oral nabilone). There were reports that pre-dosing with CBD or higher CBD content in cannabis might protect against some THC-induced deficits in verbal learning and memory. Verbal learning and memory were most often measured by means of word-list learning tasks, with several immediate and delayed recall trials and a recognition task. Attention was most often measured by means of tasks of divided attention, processing speed, rapid visual information processing, visual search, and tracking. Psychomotor function was often measured by means of tasks involving finger tapping, critical tracking, and choice reaction time. There were reports of impairments of verbal memory, of attention, and of some executive functions persisting after prolonged abstinence, but persistence or recovery across all cognitive domains remains under-researched. Psychomotor function was reportedly most affected during acute intoxication, and there was some evidence of the effect persisting in chronic users after they stopped using for a prolonged period. There was less consistent evidence for impairment of executive function (planning, reasoning, problem solving) in either occasional, moderate, or heavy users.

On the basis of their extensive review, Broyd et al. suggest that the literature on the cognitive effects of cannabis exposure is unclear as to the amount of cannabis exposure in samples and the means by which cognitive function is assessed. In studies with chronic users, it cannot be discerned whether the effects are attributable to chronic long-term exposure or to residual or cumulative acute affects resulting from the buildup of THC in tissue. Confounding factors that may affect attribution of impairment to cannabis include pre-morbid functionality and other substance abuse. Only a few large-sample prospective studies have controlled for cognitive ability assessed before initiation of cannabis use in order to address pre-morbid functionality.

Whether chronic cannabis use produces long-term impairment of cognitive abilities in individuals who are no longer using cannabis is controversial, and the literature is fraught with contradictions. Polydrug abuse and pre-existing cognitive and emotional differences between cannabis users and non-users make interpretation of the literature on studies of humans problematic. Some investigators, among them Solowij and Battisti (2008), conclude that chronic exposure to marijuana is associated with dose-related cognitive impairments—most consistently, impairments in attention and short-term working memory functions similar to the acute effects on memory. Others, including Dregan and Gulliford (2012), conclude that few if any cognitive impairments are produced by heavy cannabis use over several years. The results of two meta-analyses (Grant, Gonzalez, Carey, Natarajan, and Wolfson 2003; Schreiner and Dunn 2012) show that non-intoxicated cannabis users perform more poorly than non-users on measures of global neuropsychological function. But when the analyses were limited to thirteen studies of cannabis users with at least one month of abstinence, there was no discernible difference between cannabis users and non-users in performance on neurological tests (Schreiner and Dunn 2012). This suggests that cognitive functions may recover with prolonged abstinence.

A recent longitudinal study reported that heavy cannabis use in adolescence was associated with a decline of approximately eight IQ points from childhood to early adulthood (Meier et al. 2012). The participants were members of the Dunedin Study, a prospective study of 1,037 individuals in New Zealand who were followed from birth (in 1972 or 1973) to the age of 38. Cannabis use was ascertained in interviews at ages 18, 21, 26, 32, and 38. Neuropsychological testing was conducted at age 13, before the age of initiation of cannabis use, and again at age 38 years; therefore, a change from pre-cannabis-use to post-cannabis-use could be ascertained for each participant. Persistent heavy cannabis use (with a diagnosis of cannabis-use disorder in at least three sampling periods) was associated with broad neuropsychological decline across domains of functioning, even after controlling for years of education. This decline was most significant among those who were heavy users in adolescence and in early adulthood. Late-onset cannabis users among this sample did not show a persistent deficit after quitting, but early-onset heavy users maintained their deficit after abstinence in adulthood. More recently, however, Rogeberg (2013) argued that the apparent link between heavy cannabis use probably was an artifact attributable to important differences such as socioeconomic status between the relatively small group of heavy cannabis users ($n = 38$) and the 1,000 other participants in the study. A causal interpretation rests on the assumption that IQ trajectories are equal across the different cannabis-exposure groups in the absence of cannabis use. The risk of selection bias is reduced because of the use of each participant as his or her own control, with the result biased only if a variable both correlates with adolescent-onset cannabis use and a time-varying effect on IQ. One such variable is likely to be socioeconomic status. A simulation of this confounding interpretation reproduced the same reported associations reported by Meier et al. (2012), which suggested that the causal effects attributed to cannabis use were likely to have been overestimated and that the true size of the effect may be zero. Therefore, further analysis of this sample is called for before a claim is made that heavy cannabis use produced the IQ decline. It should also be noted that in the study by Meier (2012) the heavy cannabis users showed impairments on every cognitive test, rather than differential effects on memory (as would be expected from earlier research).

Since the Meier study (reported in 2012), two additional large prospective cohort studies have assessed the relationship between cannabis use and IQ, one in the United Kingdom (Mokrysz et al. 2016) and one in the United States (Jackson et al. 2016). In the UK birth-cohort study of 2,235 adolescents (15–16 years old) it was found that cumulative cannabis use was not associated with a lower IQ relative to non-using controls when IQ measures were compared against those taken before the teen years and when alcohol and tobacco use were removed as potential confounders (Mokrysz et al. 2016). In addition, the US birth-cohort study of 3,066 individuals between the ages of 17 and 20 found no difference in IQ from that measured at ages 9–12 between identical and non-identical twins discordant for cannabis use (Jackson et al. 2016). Although these recent studies are limited by participants reporting fewer cannabis exposures than in the study by Meier et al. (2012), there does not appear to be support for a causal link between cannabis use and long-term change in IQ. Furthermore, all studies have relied on retrospective self-reporting of cannabis use, have ignored potential residual effects of the drug on IQ test performance, and have not addressed the strain or potency of the cannabis used (Curran et al. 2016).

Before the study by Meier et al. (2012), there had been only one similar longitudinal study of the long-term effects of marijuana on cognition in participants who were exclusively cannabis users (Fried, Watkinson, and Gray 2005). Fried et al. (2005) conducted a longitudinal examination of young adults using neurocognitive tests that had been administered before the first experience with marijuana smoke. Users were defined as light (fewer than five times a week), heavy (more than five times a week), current, or former (abstinent for at least three months) cannabis users. Only the current heavy users performed worse than non-users in overall IQ, processing speed, and immediate and delayed memory tests. Former heavy marijuana smokers did not show any cognitive impairment.

A number of studies of long-term effects after an individual stops using cannabis are converging to show that cognitive impairments do not persist beyond four to six weeks after abstinence. (See Curran et al. 2016.) Chronic cannabis users showed downregulation of cortical CB_1 receptors that was correlated with years of use (Hirvonen et al. 2012), but after about four weeks of continuously monitored abstinence their CB_1-receptor density returned to control levels. This may occur after only two days (D'Souza et al. 2016). Similar findings have been reported from pre-clinical animal studies (Sim-Selley 2003).

Drawing conclusions from the human literature is challenging. Investigators use widely differing methods with different tasks and often lacking sufficient controls. Participant-selection strategies differ between studies. Across studies, cannabis is given by different routes of administration and different doses of THC are used. Sample sizes are often very small, and use of multiple drugs is common. In addition, factors such as a predisposition to substance use in general may confer greater vulnerability to cannabis-related cognitive effects. Larger-scale longitudinal studies currently underway may provide a better understanding of the effects of chronic exposure to marijuana on cognitive function independent of pre-existing differences or co-morbidity with alcohol. Therefore, our current understanding of the effects of cannabinoids on various processes involved in learning and memory relies heavily on animal models, which provide insights into the role of the endocannabinoid system in the physiology of learning and memory.

Effects of CB_1 Agonists on Learning and Memory in Non-Humans

Considerable pre-clinical animal research has evaluated how CB_1 agonists affect learning and memory. Some of these studies involve administration of THC and other global CB_1 agonists; others involve manipulations of the endocannabinoid system by administration of FAAH inhibitors (which elevate AEA and other fatty acids) and MAGL inhibitors (which elevate 2-AG when and where they are produced).

Effect of systemic administration of THC and other global CB_1 agonists

Consistent with the human literature, the effect of CB_1 agonists selectively disrupts short-term or working memory but not long-term or reference memory. The direct effect of CB_1 agonists also reduces the transfer of short-term memories into long-term memory (called *consolidation*). Several animal models of short-term or working memory versus reference memory have been developed to evaluate these different aspects of memory. One common model is delayed matching (or non-matching) to a sample task in which the animal must first learn to perform an operant task (pressing a bar to receive food reward) that represents the reference memory or long-term memory aspect of the task. Once the animal has learned to perform that operant task, it must then indicate (usually by pressing a bar) which test sample matches (or does not match) the original sample stimulus presented several seconds earlier (working memory or short-term memory). CB_1 agonists disrupt only the working-memory task or the short-term memory task, with a greater disruption the longer the delay between the stimuli. (See Mechoulam and Parker 2013.) These effects are blocked by the CB_1 antagonist rimonabant. It is important to note that these effects occur at low doses that do not interfere with the acquisition of the original reference memory of the task.

A simpler task for the assessment of short-term memory is called the *object-recognition task*. It relies on a rodent's natural preference to explore novel objects. A rat or a mouse is allowed to spontaneously explore two identical novel objects. When it is exposed to the same object at a second encounter shortly after the first exposure, investigatory behavior is considerably reduced. In the object-recognition task, administration of CB_1 agonists before the choice task impairs memory for the familiar object. Furthermore, chronic exposure to CB_1 agonists for 21 days produces a deficit in object-recognition memory even after a 28-day drug-free period (O'Shea et al. 2006), regardless of the age of pre-exposure (perinatal, adolescent, or adult).

Accurate working or short-term memory is also required for spatial-memory tasks, such as that involving an eight-arm radial maze. In that task, rats or mice must first learn which arms contain food rewards (reference memory) and must then remember which arms have already been visited in a test session (working memory) after an imposed delay. THC increases the number of working-memory errors (re-entries) at low doses, and these effects are blocked by rimonabant (Lichtman and Martin 1996). Chronic (90 days) exposure to THC enhanced the memory impairment in this task, but the impairment disappeared after 30 days of abstinence from the drug (Nakamura, da Silva, Concilio, Wilkinson, and Masur 1991). On the other hand, impairment in spatial working memory was seen in adolescent rats treated with very high escalating doses of THC (2.5–10 mg/kg) chronically for 10 days and left undisturbed for 30 days until adulthood. The working-memory deficit was also accompanied by a decrease in hippocampal dendritic spine density and length (Rubino and Parolaro 2008).

Another commonly employed task for assessing working or short-term-memory deficits is the Morris water maze, which requires animals to locate a hidden platform in a pool of water using salient visual cues surrounding the pool. The water-maze task can be used to evaluate the effect of cannabinoid agonists on reference or long-term memory (location of the platform remaining fixed across days and on trials within a day) or on working or short-term memory (location of platform is changed each day, but remains constant across trials within a day). In the water-maze task, THC disrupts working memory at doses much lower than those that disrupt reference memory. The doses that are sufficient to disrupt working memory are lower than those that produce other effects characteristic of CB_1 agonism, including anti-nociception, hypothermia, catalepsy, and hypomotility (Varvel, Hamm, Martin, and Lichtman 2001). Vaporized marijuana smoke produces a similar effect (Niyuhire, Varvel, Martin, and Lichtman 2007).

Effects of endocannabinoid manipulations on learning and memory

Exogenous administration of CB_1 agonists affects CB_1 receptors throughout the brain. On the other hand, administration of inhibitors of the degrading enzymes FAAH (elevating AEA and other fatty acids, including OEA and PEA) and MAGL (elevating 2-AG) produce effects only in brain regions activated by the manipulation, and only the CB_1 receptors within regions relevant for learning and memory are activated. Although exogenous CB_1 agonists consistently suppress working memory in these models of short-term memory, manipulations that elevate endogenous cannabinoids "where and when they are needed" do not consistently produce such an impairment. On the one hand, elevation of AEA (by FAAH inhibition), but not 2-AG (by MAGL inhibition), interfered with the consolidation of contextual conditioned fear and object-recognition memory (Busquets-Garcia et al. 2011]. On the other hand, several studies (Campolongo et al. 2009; Mazzola et al. 2009; Varvel, Wise, Niyuhire, Cravatt, and Lichtman 2007) have reported that FAAH inhibition facilitates rather than interferes with working memory. Varvel et al. (2007) reported that mice deficient in FAAH, either by genetic deletion or by pharmacological inhibition, displayed both faster acquisition and faster extinction of spatial memory tested in the Morris water maze; rimonabant reversed the effect of FAAH inhibition during both task phases. FAAH-deficient mice (with tenfold increases in brain levels of AEA) also showed improved rather than impaired performance in these tasks. Therefore, the effects of exogenously administered CB_1 agonists are not always consistent with the effects of manipulations that elevate the natural ligands for the receptors. However, FAAH inhibition elevates not only AEA but also several other fatty acids (including OEA and PEA, which are ligands for PPAR-α). Mazzola et al. (2009) found that enhanced acquisition of a passive avoidance task by the FAAH inhibitor URB597 was reversed not only by a CB_1 antagonist but also by the PPAR-α antagonist MK 886. The PPAR-α agonist WAY1463 also enhanced passive avoidance performance, and its effect was blocked by a PPAR-α antagonist (Campolongo, Roozendaal, Trezza, Cuomo, et al. 2009). Therefore, FAAH inhibition may enhance memory not only by increasing AEA but also by elevating OEA and PEA. MAGL knockout mice, with elevated levels of 2-AG, also show improved learning in an object-recognition task and in a water-maze task (Pan et al. 2011). Thus, there is evidence that both AEA and 2-AG may be cognitive enhancers under some conditions. However, simultaneous elevation of both AEA and 2-AG with a dual FAAH/MAGL inhibitor, JZL195, revealed that concomitant increases in AEA and 2-AG actually disrupt short-term spatial memory performance in a manner similar to that of THC (Wise et al. 2012).

Effects of CB_1 Antagonists on Learning and Memory in Non-Humans

Since CB_1 agonists produce deficits in short-term working memory, it may be expected that CB_1 antagonists should show enhancement of short-term memory—that is, serve as cognitive enhancers. However, the literature is replete with mixed findings. CB_1 antagonist administration produced memory enhancement in mice in an olfactory recognition task and a spatial-memory task in an eight-arm radial maze. In addition, mice genetically deficient in CB_1 receptors are able to retain memory in an object-recognition test for at least 48 hours after the first trial, whereas wild-type controls lose their capacity to retain memory after 24 hours. In contrast, studies using other paradigms, particularly delayed matching to sample, have shown no benefits of rimonabant on learning or memory. (See Mechoulam and Parker 2013.) One explanation (Varvel, Wise, and Lichtman 2009) for the mixed findings is that the temporal requirements of the task determine whether the CB_1 antagonist will facilitate performance. Studies showing enhancement of memory generally require memory processes lasting minutes or hours (e.g., object recognition; social recognition), whereas studies showing that rimonabant is ineffective generally require retention of information lasting for only seconds (e.g., delayed matching to sample), which suggests that blockade of CB_1 receptors may prolong the duration of a memory rather than facilitate learning. (See Varvel et al. 2009 for a review.)

The Role of Endocannabinoids in the Hippocampus in Learning and Memory

The decrement in working or short-term memory caused by cannabinoids is mediated by their action at the hippocampus. The hippocampus has a high density of CB_1 receptors. The detrimental effects of CB_1 agonist on working memory, but not reference memory, parallel the effects of hippocampal lesions on those two forms of memory. Intracranial administration of the CB_1 agonists directly into the hippocampus also disrupts the performance of working memory in an eight-arm radial maze, in water-maze spatial learning, and in object-recognition memory. In contrast, intrahippocampal administration of a CB_1 antagonist has been shown to disrupt memory consolidation of an inhibitory avoidance task. (See Mechoulam and Parker 2013.) Cannabinoid and the cholinergic systems have been shown to interact in the hippocampus during performance of a short-term memory task in rats (Goonawardena, Robinson, Hampson, and Riedel 2010). These effects may be mediated by cannabinoid-induced decreases in acetylcholine release in the hippocampus. Acetylcholine is also implicated in the pathophysiology of Alzheimer's Disease and other disorders associated with declined cognitive function.

Overall, the literature implicates changes in hippocampal functioning as the source of working-memory deficits produced by THC, although other brain regions are being investigated. (For a review, see Marsicano and Lafenetre 2009.) Cannabinoid receptors localized to different brain regions modulate distinct learning and memory processes, and so the role of endocannabinoids in other regions may be different from their role in the hippocampus. For instance, infusion of a CB_1 agonist into the basolateral amygdala enhanced consolidation of inhibitory avoidance learning by enhancing glucocorticoid action in that region (Campolongo, Roozendaal, Trezza, Hauer, et al. 2009), and infusion of a CB_1 antagonist into that region interfered with aversive learning. (See also Tan et al. 2011.) O'Brien et al. (2014) reported that infusion of a CB_1 antagonist into the interoceptive insular cortex (a region also implicated in addiction and nausea) or into the somatosensory area instead facilitated consolidation of object-recognition memory. The differential effects of CB_1 agonists and antagonists on different brain regions may account for the different findings reported after systemic and localized administration of cannabinoid agonists.

Memory formation in the hippocampus is believed to be mediated by long-term changes in synaptic plasticity, called *long-term potentiation* (LTP). One of the most interesting characteristics of LTP is that it causes long-term strengthening of the synapses between two neurons that are activated simultaneously. LTP relies on the activation of NMDA glutamate receptors. CB_1 agonists may impair working memory by suppressing release of glutamate in the hippocampus. The suppression of communication across two synapses, called *long-term depression* (LTD), is the opposite of LTP. Retrograde signaling by endocannabinoids results in suppression of neurotransmitter release at both excitatory (glutamatergic) and inhibitory (GABAergic) synapses in the hippocampus in a short-term manner and in a long-term manner, producing LTD,. (The latter is one of the best examples of pre-synaptic forms of long-term plasticity.) Recent evidence indicates that pre-synaptic activity coincident with CB_1-receptor activation and NMDA-receptor activation is required for some forms of endocannabinoid LTD. The long-lasting effects of LTD appear to be mediated by a CB_1-receptor-induced reduction of cAMP/PKA activity in the hippocampus (Heifets and Castillo 2009).

Results published in 2013 in the journal *Cell* showed that chronic exposure to THC reduced both hippocampal LTP and performance in hippocampus-dependent memory tasks in mice. These effects were accompanied by increased cyclooxygenase 2 (COX2) signaling and its product prostaglandin E2 (PGE2) in the hippocampus (Chen et al. 2013). Administration of a COX2 inhibitor prevented the detrimental effects of THC on LTP, on learning, and on production of PGE2. That deficits in LTP and memory were not seen in mice lacking COX2 suggests that COX2 inhibitors (e.g., ibuprofen) administered with THC may reduce the side effect of memory impairment. Recent pre-clinical findings suggest that such combined treatments may be useful in Alzheimer's Disease. Daily injection of THC for four weeks reduced both amyloid-β (a protein implicated in Alzheimer's Disease) and neurodegeneration in 5XFAD APP transgenic mice (a model of Alzheimer's Disease) whether or not the injections were accompanied by administration of a COX2 inhibitor. Therefore, adjunct COX2 inhibition did not affect this particular beneficial effect of THC. These findings suggest that COX2 inhibitors (e.g., ibuprofen) may enhance the medical utility of marijuana by reducing the side effect of memory impairment (Chen et al. 2013).

Long-Term Effects of Marijuana on Brain Morphology and on Cognitive Processes

Marijuana use has increased over the past ten years, as has the potency of THC in the marijuana. Since it is known that THC produces at least acute effects on short-term memory processes, it is possible that long-term chronic use of high-potency marijuana may produce neural changes affecting cognitive processes. However, whether brain changes result from long-term marijuana use is equivocal. Some studies report functional changes in cognition in both adults and adolescents; others report no changes. Some studies report decreased volume of sub-cortical regions; others report increased volumes in the same or different regions in chronic marijuana users (Weiland et al. 2015). Indeed, when the sizes of effects are averaged across all studies, Weiland et al. (2015) report a mean cumulative effect side of $d = -0.011$, which suggests no effect within the bounds of sampling error.

Filbey et al. (2014) collected a cross-sectional large group of chronic marijuana using adults with a wide age range, allowing for a measure of changes across lifespan. They found that chronic marijuana use was associated with complex neuroadaptive processes and that age of onset and duration of use had unique effects on these processes. There was a high co-morbidity of marijuana use, alcohol use, and tobacco use, so they divided the groups into exclusive marijuana users. They found that heavy (at least four times per week over past six months) chronic users (either co-morbid or exclusively marijuana users) had lower orbitofrontal cortex gray-matter volumes relative to non-using controls. The orbitofrontal cortex is a region in the reward network and implicated in addiction and in decision making. There were no differences in other regions, but the effects in the orbitofrontal cortex were even greater in exclusively marijuana users than in co-morbid users. However, as Filbey et al. (2014) noted, the cross-sectional nature of their study cannot assess whether these reductions are the cause or the consequence of marijuana use—that is, there could be pre-existing differences that lead to marijuana use in the first place. Longitudinal studies in humans are needed to address the causality of these neural abnormalities. Indeed, the results of a four-year longitudinal study by Cheetham et al. (2012) showed that individuals with smaller orbitofrontal cortex volumes at age 12 were more likely to initiate cannabis use (as well as tobacco and alcohol use) by age 16. Therefore, there is evidence that the structural abnormality in the orbito-frontal cortex reported by Filbey et al. (2014) may have been present before cannabis use began. The orbitofrontal cortex underlies inhibitory and decision-making processes that might influence risk for early canna-bis use. Further studies are needed to determine whether any potential changes in brain function or structure that may be present in cannabis users revert back to normal after protracted abstinence from marijuana use. The small literature suggests that cognitive alterations and CB_1-receptor downregulation in regular marijuana users may return to nor-mal values as a result of a neuroadaptive phenomenon occurring after a period of abstinence.

Conclusion

Evidence suggests that the acute effects of cannabis on memory depend in part on the type of cannabis that is used. Smoking cannabis with higher levels of CBD protected users against the acute memory-impairing effects of THC (Morgan et al. 2010).

Global activation of CB_1 receptors produces actions on the hippocam-pal neurons that result in interference with short-term memory and in interference with consolidation of memories that are currently being pro-cessed. However, such CB_1-receptor activation does not impair recall of previously established memories. These effects on the formation of new memories would be particularly problematic for adolescent use of can-nabis. It still isn't clear whether any permanent changes in learning and memory processes result from early exposure to cannabis in now-abstinent individuals. Insofar as most cannabis users also use alcohol, which is known to be a cognitively impairing drug (Curran et al. 2016), it is not clear how repeated use of cannabis causally affects the human adolescent and adult brain. The recent findings that administration of COX-2 inhibitors (e.g., ibuprofen) or CBD with THC may prevent memory-impairing effects of THC are promising avenues for further research.

6

Cannabinoids, Reward, and Addiction

The euphoric effects of cannabis became known to the Western world through the writings of Baudelaire, Gautier, and Moreau in the middle of the nineteenth century. During monthly meetings of the Club de Hachichins at the Hotel Lauzun in Paris's Latin Quarter, Moreau dispensed dawamesk (a mixture of hashish, cinnamon, cloves, nutmeg, pistachio, sugar, orange juice, butter, and cantharides) to eminent people who had assembled to ingest the drug. "There are two modes of existence—two modes of life—given to man," Moreau mused. "The first one results from our communication with the external world, with the universe. The second one is but the reflection of the self and is fed from its own distinct internal sources. The dream is an in-between land where the external life ends and the internal life begins." With the aid of hashish, he felt that anyone could enter this in-between land at will (Abel 1980). As Moreau studied hashish, he noted a relationship between the amount of the drug taken and its effects. A small dose produced a sense of euphoria and calmness. As the dose increased, attention wandered, ideas appeared at random, minutes seemed like hours, thoughts rushed together, and sensory acuity increased. As the dose increased further, dreams began to flood the brain, like hallucinations of insanity (ibid.).

Can cannabis can be considered an abused drug? Is it addictive? How does it produce an effect on reward centers of the brain? Does it change the morphology of these centers of reward in regular users? How does it influence relapse to self-administration of other drugs of abuse? What are the interactions between cannabinoids and opiates? How does cannabis influence sexual motivation and desire?

The Abuse Potential of Cannabinoids

Humans have an urge to alter their state of consciousness, and the use of psychoactive drugs is a means of achieving that goal. Understanding how such drugs act on the brain and how their use can turn to abuse and addiction is important for the health of individuals and for society at large. Methamphetamine, heroin, nicotine, and alcohol clearly have the potential to produce addiction in humans. Whether or not cannabis is addictive has been more controversial.

The term "addiction" generally refers to pathological drug taking that ranges from mild to severe. At first, intake of a drug gradually escalates from sporadic recreational use to more sustained and regular intake; later, discontinuing use of the drug becomes difficult. The later phase, generally referred to as "addiction," is characterized by loss of control of drug use. The individual cannot control the amount of the drug taken, spends more time in drug-seeking and drug-taking activities, and cannot stop using the drug despite adverse consequences. When discussing addiction, the fifth edition of the American Psychiatric Association's *Diagnostic and Statistical Manual of Mental Disorders* (DSM) uses the term "substance-related disorder" rather than "addiction," partly because "addiction" is commonly used to refer to only the severe final stage of a substance-use disorder. This distinction recognizes that there is a continuum in the severity of the disorders related to drug use. "Use disorder" is meant to indicate that a person can have a problem with substances that does not necessarily coincide with the extreme case in which behavior is the kind of uncontrolled compulsive drug seeking typically seen in "addiction." The range of symptoms and consequences seen in the clinic is quite large, and effective markers across individuals are not necessarily clear, but the *DSM*'s fifth edition provides a means of designating severity level (mild, moderate, severe) by summing the number of diagnostic criteria displayed (Piomelli et al. 2016).

Cannabis-use disorder
Despite earlier beliefs that cannabis is not addictive, clinical and experimental research conducted over the past twenty years has demonstrated that people may become dependent upon cannabis, in what is called *cannabis use disorder*. However, most agree that cannabis-use disorder does not result in the same extreme levels of behavior -apparent with addiction to other drugs of abuse. One of the top clinical researchers in this field, Alan J. Budney, has described cannabis-use disorder as follows:

The features of cannabis use disorder are the same as the features of all substance use disorders according to the *DSM* 5th edition. There is a list of 11 criteria, and they encompass the range of signs and symptoms that can be experienced, including physiological signs like tolerance and withdrawal to cannabis, continuing to use cannabis despite the person knowing he or she has problems being caused by cannabis, and recurrent use in situations that might be hazardous or dangerous like driving a car while high on cannabis. Other signs include using to such excess that cannabis seems to take over one's life, and healthy behaviors like work and recreation and positive relationships are harmed, ignored, or greatly reduced. Experiencing strong and frequent cravings to use, and using more cannabis than one plans to use or for a much longer time than was planned are also common features of the disorder. Last, people who develop problems with cannabis may have repeated desire to cut down or quit, but end up going back to using the same amount or more. Some scientists have tried to identify, statistically, hallmark signs or symptoms of cannabis use disorder that differentiate it from other substance use disorders, but they have not been successful. Essentially, cannabis use disorder manifests in the same way as other substance use disorders, but the difference may be in the magnitude of severity of each of the signs and symptoms that are expressed. One example of this is the difference in severity of withdrawal symptoms experienced by those who abruptly stop using opiates versus cannabis. Heavy cannabis users who stop experience withdrawal symptoms that may be somewhat similar to tobacco withdrawal symptoms, but they do not approach the severity nor have the clinical implications of the withdrawal experience by many opiate users. (Piomelli et al. 2016, p. 48)

Abstinence from marijuana after daily use can produce a withdrawal syndrome that appears within the first one or two days after discontinuation; its effects reach their peak after 2–6 days, and most symptoms (e.g., irritability, anxiety, muscle pain, chills, nightmares, insomnia, headache, decreased appetite) resolve within a week or two (Bonnet, Specka, Stratmann, Ochwadt, and Scherbaum 2014). Oral administration of THC reduces these withdrawal symptoms (Haney et al. 2004). Indeed, about 24 percent of patients now entering treatment for substance abuse have a diagnosis of cannabis-use disorder (Substance Abuse and Mental Health Services Administration 2013).

There are no approved pharmacotherapies for managing the symptoms of withdrawal from cannabis (Allsop et al. 2014). Anti-depressants and mood stabilizers have been used, but with only limited benefit. Agonist-substitution therapy may be more promising; indeed, dronabinol and nabilone may be more promising. A placebo-controlled, within-subject clinical study of daily cannabis smokers not seeking treatment reported that 6-mg and 8-mg daily doses of nabilone decreased cannabis relapse and reversed cannabis-related irritability and disruptions of sleep and food intake (Haney et al. 2013). Neither dose of nabilone increased self ratings of drug "liking, an index of psychoactive euphoria effects, but the highest dose worsened psychomotor performance slightly. In addition there is evidence that nabiximols (Sativex 2.7 mg THC: 2.5 mg CBD/spray) may show promise as an agonist substitute treatment for cannabis-use disorder. A randomized, double-blind, placebo-controlled inpatient clinical study found that nabiximols attenuated symptoms of cannabis withdrawal and improved patient retention in treatment; however, placebo was as effective as nabiximols in promoting long-term reductions in cannabis use at follow-up (Allsop 2014). Nabiximols treatment reduced the overall severity of cannabis withdrawal symptoms relative to placebo, including effects on irritability, depression, and craving and more limited effects on sleep disturbance, anxiety, appetite loss, and restlessness (Allsop et al. 2014). A recent double-blind crossover study compared the abuse liability of nabiximols with oral THC (dronabinol) using standard measures of drug discrimination and drug liking (Schoedel et al. 2011). At moderate doses, nabiximols did not produce significant adverse cognitive or psychomotor side effects and showed lower abuse potential than dronabinol; however, at higher doses both medications exhibited abuse potential (defined as self-reported drug liking relative to placebo).

One reason cannabis was considered not addictive for many years was the inability of experimental animal-based research to demonstrate the rewarding effects of THC. That is, for a long time scientists had great difficulty demonstrating that animals in experiments would self-administer cannabis or THC-type compounds. Much of the hallmark work done in the early days of addiction research was done in animal labs. Thus, the difficulty of demonstrating cannabis self-administration or withdrawal in animals made it difficult to claim that cannabis was an addictive substance, as opioids and cocaine were said to be. In addition, the fact that many users of cannabis did not develop problems or experience symptoms of withdrawal led to the perception that cannabis is not addictive. Finally, the potency of THC in cannabis has been rising, which may account (partially, at least) for the fact that the number of individuals diagnosed with cannabis-use disorder has increased by 8–15 percent in the past 15 years (Piomelli et al. 2016; Substance Abuse and Mental Health Services Administration 2013). A very interesting recent finding indicated that, when intoxicated, human smokers of high-CBD:THC strains showed reduced attentional bias to drug and food stimuli relative to smokers of low CBD:THC strains. Those smoking high-CBD:THC strains also showed lower self-rated liking of cannabis stimuli (a measure of abuse potential in humans). These findings suggest that CBD has potential as a treatment for cannabis-use disorder (Morgan, Freeman, Schafer, and Curran 2010).

Are cannabinoids rewarding drugs in animal models?

A wide range of drugs can be shown to act as reinforcers in rodents and monkeys, the relative reinforcement potential and pattern of drug taking closely resembling those seen in humans. In addition, antipsychotic drugs and other drugs not self-administered by animals lack reinforcing properties or may even be aversive to humans. The two best-characterized animal models of the rewarding effects of drugs are the drug self-administration procedure (figure 6.1) and the conditioned place preference paradigm (figure 6.2). In the drug self-administration

Figure 6.1
A self-administration apparatus. Rats implanted with intravenous catheters are trained to self-administer drugs by pressing a lever. The rat is first trained on an FR-1 schedule or an FR-5 Schedule (that is, one or five lever presses, respectively, are required for an infusion of the drug). Animals titrate the amount of the drug to maintain an optimal level if it is rewarding.

paradigm, an animal must learn to press a lever in order to receive an intravenous injection of a drug. The animal is usually trained on a Fixed Ratio-1 (FR-1) schedule or a Fixed Ratio-5 (FR-5) schedule of reinforcement. In an FR-1 schedule, the animal receives an intravenous infusion of the test drug every time it presses the lever. An FR-5 schedule requires that the animal press the lever five times before it receives an infusion of the drug. Animals quickly learn to "titrate" the amount of the drug they receive to maintain an optimal level; that is, they learn to regulate the dose. If the dose of the drug infused is decreased below the training dose, the animal will self-administer more infusions; if the dose of the drug infused in increased above the training dose, it will self-administer fewer

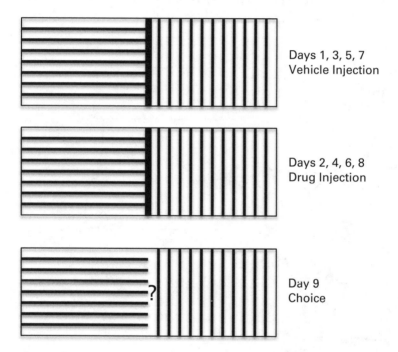

Days 1, 3, 5, 7
Vehicle Injection

Days 2, 4, 6, 8
Drug Injection

Day 9
Choice

Figure 6.2
The procedure for place conditioning. Rats or mice are injected with the vehicle in which the drug is prepared on days 1, 3, 5, and 7 and placed in the chamber with the horizontal stripes. On days 2, 4, 6, and 8 the same rats or mice are injected with a drug and placed in the chamber with the vertical stripes. The chambers and the order of the conditioning trials are counterbalanced among the rats. On the test day, the barrier between the two chambers is removed and the drug-free rodents are given a choice between the drug-paired chamber and the vehicle-paired chamber. If they spend more time in the drug-paired chamber than in the vehicle-paired chamber, the drug is considered rewarding; if they spend less time in the drug-paired chamber than in the vehicle-paired chamber, the drug is considered aversive.

infusions. In addition, through a process of classical conditioning, stimuli (such as a light) that are paired with infusion of a drug acquire conditioned reinforcing properties. These stimuli can be subsequently presented to the animal to elicit "drug-seeking" behavior, which means responding in anticipation of receiving the drug (craving). This paradigm is useful in evaluating putative medicines for controlling drug craving which promotes relapse to drug taking in humans.

Early research suggested that neither monkeys nor rodents self-administered THC. However, the use of more sensitive techniques has resulted in more reliable evidence that very low doses of THC (but not high doses), as well as anandamide and 2-AG, are self-administered by monkeys (Justinova et al. 2011) with or without a history of cannabinoid self-administration. Although the FAAH inhibitor URB597 is not self-administered by monkeys (which suggests that it does not have abuse potential), it does make anandamide more rewarding in the self-administration paradigm in monkeys (Justinova et al. 2008).

In the conditioned place preference paradigm (figure 6.2), a rodent is injected with a drug and placed in a distinctive chamber on one day and is injected with the vehicle in which the drug is prepared and placed in a different distinctive chamber on another day, with the order counterbalanced. This procedure is usually continued until the rat has experienced three or four pairings of the drug with the chamber. Then the animal, when drug free, is given a choice between the two chambers. If the drug is rewarding, the rat will return to the chamber previously paired with the drug. If the drug is aversive, the rat will avoid the chamber paired with the drug. Although abused drugs such as cocaine, amphetamine, morphine, and heroin consistently produce a conditioned place preference, conflicting findings have been reported with THC in this model. It appears that THC produces a place preference in rodents at very low doses (Lepore, Vorel, Lowinson, and Gardiner 1995) and in rodents that have had experience with THC (Valjent and Maldonado 2000). Otherwise, at higher doses and in naive rodents, THC is aversive, probably as a result of its anxiogenic effects (Parker and Gillies 1995).

In contrast to exogenous cannabinoid agonists, pharmacological enhancement of endocannabinoid levels by FAAH inhibition or MAGL inhibition generally does not produce rewarding effects. In most animal studies reported to date, neither FAAH inhibitors nor MAGL inhibitors support operant self-administration or produce conditioned place preference in rats or mice. In addition, FAAH or MAGL inhibitors (or exogenously administered AEA or 2-AG) do not produce THC-like discriminative stimulus effects in rats or mice. On the other hand, exogenous AEA and 2-AG are both self-administered by squirrel monkeys and produce rewarding effects in rats when co-administered with FAAH inhibition (Justinova et al 2008; Parsons and Hurd 2015). Concurrent FAAH and MAGL inhibition in mice produces THC-like discriminative and behavioral effects (Wise et al. 2012). These findings suggest that the endocannabinoid system has to be robustly activated in order to evoke rewarding effects.

The Action of Cannabinoid on the Reward System of the Brain

Drugs of abuse, including cocaine, amphetamine, opiates, and alcohol, engage the mesolimbic dopamine (DA) system (Wise 2004). (See figure 6.3.) There is evidence that cannabinoids, too, activate this reward system of the brain. The primary role of endocannabinoids is to regulate the release of other neurotransmitters, including GABA, glutamate, and dopamine. However, the effect on DA is not direct; it is indirectly regulated by action on GABA and glutamate. As is illustrated in figure 6.3, DA afferent axons originate from the ventral tegmental area (VTA) and project to the nucleus accumbens (NAc). The NAc consists primarily of inhibitory GABA neurons, including local interneurons and medium spiny neurons that project to the VTA. The NAc receives excitatory glutamatergic input from limbic regions, including the prefrontal cortex and the amygdala. This excitatory input results in release of GABA within the VTA, thereby inhibiting the release of DA in the NAc. Endocannabinoids indirectly control the release of DA by their inhibitory action on the release of glutamate in the NAc (which reduces GABA release in the VTA), as well as by their direct inhibition of GABA release in the VTA. This results in disinhibition of DA transmission from the VTA to the NAc (Lupica, Riegel, and Hoffman 2004; Riegel and Lupica 2004). Therefore, the final effect of endocannabinoids on the modulation of DA activity depends on the functional balance between the inhibitory GABAergic and excitatory glutamatergic inputs to the VTA, the latter predominating (Maldonado, Valverde, and Berrendero 2006). Since all addictive drugs display the ability to increase DA in the NAc (e.g., Wise 2004), this probably is the mechanism underlying the rewarding effects of CB_1 agonists.

Positron-emission tomography studies indicate a small increase in dopamine release in the striatum after administration of THC to humans (Bossong et al. 2015; Bossong et al. 2009), although the increase is much

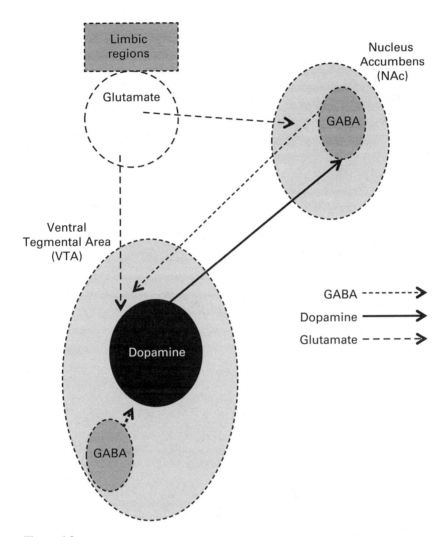

Figure 6.3
Interaction between the endocannabinoid and dopaminergic systems in the mesolimbic reward circuit. The ventral tegmental area (VTA) produces dopamine (DA) and projects afferents to the nucleus accumbens (NAc). The release of DA in that region produces the rewarding effects of natural reinforcers and drugs of abuse. Treatments that reduce the amount of dopamine released interfere with reward. Endocannabinoid regulation of DA occurs in the NAc when glutamate (GLU) projections from the limbic regions of the prefrontal cortex and the amygdala trigger the production of endocannabinoids, which reduce further release. This reduction of glutamate release reduces the release of GABA from the NAc, which feeds back to the VTA, thereby disinhibiting (increasing) the release of DA from the VTA to the NAc. The activation of CB_1 present on axon terminals of GABAergic neurons in the VTA, inhibits GABA transmission and removes this inhibitory input on DA neurons leads to an increasing in the firing of these cells (Riegel and Lupica 2004).

smaller than those observed after administration of other recreational drugs (Curran et al. 2016). On the other hand, other studies have shown no change in dopamine transmission (Barkus et al. 2011; Stokes et al. 2009). Furthermore, Urban et al. (2012) found that mild to moderate cannabis dependence was not associated with striatal DA alterations, which was consistent with an earlier report (Sevy 2008). More recently, cannabis users were reported to show a *reduction* in dopamine release (Bloomfield et al. 2014). The available human data provide only weak support for dopaminergic involvement in cannabis addiction in humans (Curran et al. 2016). In animal models, the conditions under which cannabinoid drugs have rewarding effects are much more restricted than those for other drugs of abuse, such as cocaine and heroin. However, when cannabinoid drugs produce reward-related behavior, similar brain structures are involved. (For an excellent recent review, see Parsons and Hurd 2015.)

Does THC Produce Changes in the Morphology of the Brain's Reward Systems?

If THC activates the dopamine reward system, chronic exposure may be expected to produce changes in the morphology of structures activated by that system. Indeed, previous work from our laboratory (Kolb, Gorny, Limebeer, and Parker 2006) found structural changes (dendritic arborization) in brain regions related to reward/aversion implicated in addiction, including the nucleus accumbens, after chronic exposure to THC in rats. However, how chronic cannabis use may modify brain morphometry in these regions in humans is not known. A recent report (Gilman et al. 2014) suggests that adolescents using marijuana more than once a week have structural changes in the gray matter (cell bodies) of the NAc and the amygdala—regions implicated in reward. That is, as in the rat study by Kolb et al. (2012), recreational users had a larger left NAc and a larger amygdala than non-users.

However, the results reported by Gilman et al. (2014) may have been confounded by differences between the marijuana users and the non-users in the use of alcohol. Indeed, a subsequent report (Weiland et al. 2015) indicates that when this difference is carefully controlled with matched groups, no structural difference in white or gray matter are found between adolescent or adult daily users of marijuana and non-users. Weiland et al. suggest that many previous studies have similar flaws. Unlike marijuana use, alcohol use has unequivocally been associated with structural brain differences and with deficits in cognition in both adults and adolescents. (See, e.g., Sullivan 2007.)

Another difference between marijuana users and non-users may be that users are more willing than non-users to engage in high-risk illegal behavior. Weiland et al. (2015) controlled for this confounding variable by equating both users and non-users on prior engagement in illegal behaviors. Their adolescent participants were recruited through juvenile justice services in Albuquerque as part of a larger study of adolescent risk behavior. Both the users and the non-users had engaged in risky behaviors, which controlled for this potential pre-existing difference. Indeed, these two groups did not differ on measures of impulsivity or sensation seeking. Thus, any difference that might be found between the two groups would be attributable to marijuana use, not to personality traits. With these careful controls, daily marijuana users did not differ from non-using controls in brain morphology. Carefully controlled longitudinal studies are necessary to ensure that the brain changes attributed to marijuana use are not confounded by alcohol co-administration or by pre-existing differences between marijuana users and non-users.

Cannabinoids and Relapse

One of the major hindrances to treatment of addiction is the high rate of relapse after abstinence from the addicting drug due to drug craving (that is, intense desire for the drug). Exposure to drug-associated stimuli, exposure to the drug itself (drug priming), and stress can precipitate drug craving and relapse in humans. The brain circuits involved in drug craving include projections from the ventral tegmental area to the nucleus accumbens and the prefrontal cortex. Repeated activation of this circuit results in the drug-associated stimuli, producing craving for the drug itself. In humans, fMRI studies have shown that cues associated with marijuana use activate the reward neurocircuitry associated with addiction (Filbey, Schacht, Myers, Chavez, and Hutchison 2009), as well as activating self-reported craving. Furthermore, alterations in the CB_1-receptor gene and in the FAAH gene have been shown to enhance fMRI activity in reward-related areas of the brain during exposure to marijuana cues (Filbey et al. 2009).

That CB_1-receptor antagonism (or inverse agonism) interferes with drug- and cue-induced relapse in animal models of self-administration and conditioned place preference learning (Serrano and Parsons 2011) is of considerable therapeutic importance. In the self-administration paradigm, the animal is first trained to self-administer the drug in the presence of a light or some other distinctive cue. Then the self-administration behavior is extinguished by allowing the animal to press the lever in the absence of infusion of the drug. After this extinction training, the potential of the drug itself (drug priming) or a cue associated with the drug to produce relapse or "drug-seeking behavior" is assessed. Such drug-seeking behavior contrasts with actual drug-taking behavior during the self-administration session. Administration of the CB_1 antagonist rimonabant prevents drug-associated cues and drug primes from producing relapse in rats and mice. Recent evidence suggests that rimonabant is more effective in interfering with drug-seeking behavior (craving leading to relapse) than drug-taking behavior during the self-administration session. Rimonabant has been shown to block drug seeking evoked by cues paired with cocaine, heroin, methamphetamine, and nicotine and drug seeking evoked by priming injections of each of these abused drugs. Therefore, blockade (or inverse agonism) of the CB_1 receptor interferes generally with drug-seeking behavior produced by drug-paired cues or by the drug itself. (For a review see Parsons and Hurd 2015.)

Because CB_1 antagonism interferes with drug-seeking behavior, considerable evidence also suggests that cannabinoid agonists reinstate heroin-seeking, cocaine-seeking, nicotine-seeking, and alcohol-seeking behavior, which suggests that cannabinoid signaling is involved in drug-seeking behavior in general, regardless of the nature of the drug. (See Parsons and Hurd 2015.) Enhancement of 2-AG levels by treatment with the MAGL inhibitor JZL184 also has been reported to potentiate relapse elicited by presentation of a nicotine-associated cue (Trigo and Le Foll 2015). Indeed, JZL184 delivered to the VTA has been reported to enhance dopamine in the NAc when assessed during drug-reward-directed behavior (Oleson et al. 2012).

Although considerable evidence indicates that agonism of the CB_1 receptor promotes cue-induced and drug-induced relapse, there is a growing literature suggesting that FAAH inhibition actually prevents nicotine reward and relapse to nicotine seeking in pre-clinical models (Forget, Coen, and Le Foll 2009; Forget, Guranda, Gamaleddin, Goldberg, and Le Foll 2016; Justinova et al. 2015; Scherma et al. 2008). Indeed, URB597, a selective FAAH inhibitor, has been shown to reverse nicotine-induced release of extracellular dopamine in the NAc shell (Scherma et al. 2008) and nicotine-induced excitation of dopamine cells in the VTA (Melis et al. 2008), perhaps by reversing nicotine-induced inhibition of GABAergic medium spiny neurons in the NAc shell. Since FAAH inhibition not only prevents the degradation of AEA (which acts on CB_1 receptors) but also oleoylethanalamide (OEA) and palmitoylethanalamide (PEA) (which act on peroxisome proliferator-activated receptor-α [PPAR-α] receptors), it is not clear which mechanism prevents nicotine reward and nicotine seeking in these studies. There is some evidence for both mechanisms. Selective PPAR-α agonists also counteract the reinstatement of nicotine seeking in rats and monkeys (Mascia et al. 2011). Justinova et al. (2015) found that URB597-induced reduction of nicotine self-administration and nicotine-priming-induced reinstatement (but not cue-induced relapse) of nicotine seeking in monkeys were reversed by a PPAR-α antagonist; however, Forget et al. (2016) found that the reduction in cue-induced relapse to nicotine seeking induced by URB597 was selectively reversed by a CB_1-receptor antagonist (at a low dose that did not affect reinstatement on its own). The role of 2-AG in relapse to self-administer nicotine had recently been reported to be the opposite of the role of AEA or other fatty acids. Indeed, it may actually potentiate relapse to nicotine triggered by re-exposure to nicotine-associated cues. Trigo and Le Foll (2015) found that the MAGL inhibitor JZL184, which selectively elevates 2-AG, has no effect on food taking, nicotine taking, or motivation for taking nicotine; instead, it selectively enhanced cue-induced relapse to nicotine without producing relapse in the absence of the nicotine-associated cue. This suggests that AEA and 2-AG may oppose one another in the regulation of relapse. Because FAAH inhibition may be useful in treating addiction, it will be important to extend these findings to other FAAH inhibitors currently being developed.

Rimonabant showed great promise as an anti-relapse treatment; however, as will be described in more detail in chapter 7, it was removed from the European market as a treatment for obesity because of psychiatric side effects of anxiety and suicidal thoughts. The generality of the effects of cannabinoids on motivational processes may explain these undesirable side effects. Since rimonabant acts not only as a CB_1 antagonist but also as a CB_1 inverse agonist, the relapse-preventing properties, and potentially the adverse side effects, may also be mediated by its inverse cannabimimetic effects, which are opposite in direction from those produced by cannabinoid-receptor agonists (Pertwee 2005). Recent evidence suggests that at least some adverse side effects of CB_1-receptor antagonists or inverse agonists seen in clinical trials involving nausea and anxiety may reflect their inverse-agonist properties (Bergman et al. 2008; Sink, Segovia, Collins, et al. 2010; Sink, Segovia, Sink, et al. 2010). It will be interesting to evaluate the potential of more recently developed

CB$_1$-receptor neutral antagonists (without inverse agonist effects), such as AM4113, to prevent drug-seeking behavior.

Activation of the CB$_2$ receptor, unlike activation of the CB$_1$ receptor, does not produce a psychoactive effect. Therefore, there is interest in recent reports that selective CB$_2$-receptor agonists were shown to inhibit intravenous self-administration of cocaine, cocaine-enhanced locomotion, and cocaine-enhanced release of DA in the NAc in wild-type and CB$_1$-receptor knockout mice but not in CB$_2$ knockout mice; this effect was blocked by a selective CB$_2$-receptor antagonist. These findings suggest that brain CB$_2$ receptors also modulate cocaine's effects (Xi et al. 2011) and provide further evidence that the CB$_2$ receptor seems to have general protective properties (Pacher and Mechoulam 2011).

Cannabidiol, the primary non-psychoactive compound in marijuana, has also been reported to attenuate cue-induced reinstatement of heroin seeking and to reverse disturbances of glutamatergic and endocannabinoid systems in the NAc produced by heroin seeking (Ren, Whittard, Higuera-Matas, Morris, and Hurd 2009). Apparently, in addition to the many other ailments that cannabidiol ameliorates (Mechoulam, Parker, and Gallily 2002), it may also be a potential treatment for heroin craving and relapse.

Interactions between Cannabinoid and Opiates

Heroin is used by only a small percentage of the population. However, non-medical use of prescription opioids is now becoming more prevalent. For instance, a 2009 study in the US reported that 6.2 million individuals were recent non-medical users of prescription opioids (Scavone, Sterling, and Van Bockstaele 2013). As many as 10 percent of high school seniors used prescription opioids for non-medical purposes in 2009—nearly as many as used marijuana. Opioid-related deaths rose by more than 300 percent between 1999 and 2006. Clearly opiate use is a major health problem in North America.

Considerable pre-clinical animal research demonstrates a clear interaction between cannabinoids and opiates at a number of levels within the cell, including direct receptor associations, alterations in endogenous opiate release, and post-receptor interactions via shared signal-transduction pathways. Various studies have demonstrated cross-tolerance, mutual potentiation, and receptor cross-talk between the mu-opiate receptor and the CB$_1$ receptor. Drugs that target the cannabinoid system often affect the opioid system in tandem. Synergistic effects of cannabinoid-opiate drugs have been reported. (See Scavone et al. 2013 for a review.) For instance, sub-threshold doses of cannabinoids and morphine reduce pain in animals and humans. Furthermore, cannabinoids have been show to exert cross-antagonism within the opioid system; that is, rimonabant can substitute for the opiate antagonist, thereby precipitating morphine withdrawal (Vigano et al. 2005). Inhibition of cannabinoid-receptor signaling by rimonabant during chronic opioid exposure reduces opioid withdrawal signs (Rubino, Massi, Vigano, Fuzio, and Parolaro 2000), and CB_1 antagonism also reduces a naloxone-precipitated morphine withdrawal induce place aversion (Wills et al 2014) by its action on the central nucleus of the amygdala (Wills et al. 2016; see also Wills and Parker 2016).

The overlapping neuroanatomical distribution, convergent neurochemical mechanisms, and comparable functional neurobiological properties of cannabinoid and opiate systems has led to the suggestion that cannabinoids could substitute for opioids to potentially alleviate withdrawal symptoms with opioid abstinence (Deroches et al. 2010). Marijuana is commonly used by opiate addicts, and there is some suggestion that it may reduce the amount of opiates used (Scavonne et al. 2013). In an opiate-dependent population, cannabis users reported spending less on opiates (mean = \$85/day) than non-cannabis users (mean = \$126/day), which suggested that they used smaller quantities of opiates.

In the past twenty years, drug overdose has become the leading cause of injury death in the United States. In 2011, 55 percent of drug-overdose deaths were related to prescription medications, and 75 percent of those deaths involved opiate painkillers. However, opiate-related deaths decreased by approximately 25 percent in 13 states in the six years after medical marijuana was legalized in those states (Bachhuber, Saloner, Cunningham, and Barry 2014). In contrast, Bachhuber et al. did not find evidence that states that had passed laws legalizing medical cannabis had different overdose mortality rates in years before the laws were passed. The implication is that medical marijuana laws, when implemented, may represent a promising approach for reducing non-intentional deaths related to opioid analgesics.

Because of the clear evidence of cannabinoid-opiate interactions, several studies have focused on the impact of cannabis exposure during treatment for opiate dependence. Although effective treatments for opiate dependence are available, the rate of dropout from treatment and the rate of relapse are high. Naltrexone is an opiate antagonist that has been used as a treatment for opiate dependence, but its effectiveness has been severely limited by poor adherence. However, Raby et al. (2009) report that opioid-dependent patients who used cannabis intermittently during naltrexone treatment showed better retention than patients with either heavy cannabis use or no cannabis use. The beneficial effect of cannabinoid agonism early in the course of naltrexone treatment suggests that it may be alleviating the withdrawal effects produced by naltrexone.

Methadone maintenance treatment is a pharmacotherapy used to eliminate illicit opiate use and reduce associated risk behaviors. The initiation of this treatment requires slow titration to avoid risk of overmedication. The initial period of dose stabilization is particular vulnerable to relapse because the patient is undergoing some symptoms of opiate withdrawal. Marijuana is commonly used in combination with heroin and continues to be used during treatment of opiate dependence (Scavone et al. 2013). Indeed, because of the interaction of cannabinoids and the opioid system, a synergy occurs between cannabinoids and opiates when they are administered concurrently. Marijuana use was recently found to increase during the early phases of dose titration in methadone maintenance treatment, and to decrease significantly after dose stabilization (Scavone et al. 2013). Ratings of opiate withdrawal decreased in methadone-maintenance patients who used marijuana during stabilization. Despite the accumulation of pre-clinical data suggesting that the benefits of cannabinoids may outweigh the risks in certain severe medical conditions, political conflict over the legalization of medicinal marijuana continues to preclude a smooth transition of these pre-clinical studies into clinical trial testing.

Cannabinoids and Sexual Behavior

Throughout history cannabis has been anecdotally described as an aphrodisiac, yet other reports suggest that it interferes with sexual behavior. (See, e.g., Indian Hemp Drugs Commission 1894, cited in Gorzalka, Hill, and Chang 2010.) The distribution of CB$_1$ receptors suggests that cannabinoids may affect sexual activity through direct effects within the hypothalamus regulating physiological and endocrinological underpinnings of sexual activity, or indirectly by modulating motor activity, anxiety, or reward mechanisms. There have been very few scientific studies of the effects of cannabis on women's sexual behavior, and most of those have been self-report studies. However, a review of these studies reveals that moderate cannabis consumption has a positive effect on female sexuality in two areas: sexual desire and sexual functioning (sexual satisfaction, pleasure, quality of orgasm).

Whereas the reports of the effects of cannabis on female sexuality are quite consistent, the reports of its effects on male sexuality are much less so. Collectively, the self-report studies with males suggest that cannabis use facilitates sexual desire while hindering erectile functioning (Gorzalka et al. 2010). The direct and indirect effects of marijuana on male and female sexual functioning are not fully understood.

Some have suggested that cannabis exerts its positive effect on sexual functioning by increasing tactile sensitivity; however, marijuana has no effect on touch sensitivity in non-sexual situations. Others have suggested that the positive effects of cannabis on sexual pleasure are related to slowing of temporal perception or to enhancement of attention, or that it is simply a placebo effect. Cannabis may raise androgen levels in women (van Anders, Chernick, Chernick, Hampson, and Fisher 2005), and there is a body of research demonstrating a link between sexual arousal and androgens in women. In men, all studies that have documented a decrease in testosterone levels after cannabis use have also found that the measured testosterone level in the users were still within the normal range. Therefore, an effect of THC on testosterone levels is not likely to explain the effect of cannabis on sexual behavior (Gorzalka et al. 2010). At present there is not enough evidence to explain marijuana's influence on sexual function in men or in women. To date, objective measurement of the effects of cannabis on human sexual functioning has not been reported.

Because of the inherent flaws of self-report data, it is important to evaluate the effect of cannabinoids on sexual behavior in non-human species. Data suggest that cannabinoids affect sexual behavior by acting centrally, specifically in the hypothalamus. There have been some reports of THC's having detrimental effects and other reports of its having beneficial effects on sexual receptivity in females. In males, however, most studies have found an inhibitory effect on sexual motivation and on erectile functioning, and temporary reductions in testosterone levels.

The effects of cannabis intake on sexual behavior and arousal appear to depend on dose in both men and women. Overall, studies suggest that cannabis facilitates sexual desire while hindering erectile function in males. In contrast, in females, at low THC doses, cannabis consumption has positive effects on both sexual desire and sexual function (Gorzalka et al. 2010). Animal studies confirm the human reports; for example, the dose-response curve for cannabinoids and sexual behavior in rats is bell-shaped.

Conclusion

Do CB_1 agonists produce rewarding effects in human and non-human animals? Most of the evidence suggests that they do, but not to the same extent as opiates or stimulants. The effects of THC and exposure to other CB_1 agonists in animal studies are to moderately elevate dopamine from afferent neurons originating in the VTA being released into the NAc. This is produced by its action on glutamate and GABA neurons, ultimately preventing the inhibition of the release of dopamine in the VTA. The current evidence suggests that chronic activation of this dopamine-mediated action does not produce morphological changes in the brains of humans who use marijuana regularly. The pre-clinical rodent work on relapse suggests that exposure to CB_1 agonists can produce relapse to self-administration of other drugs of abuse and that CB_1 antagonists can prevent both drug-priming-induced relapse and relapse triggered by re-exposure to drug-paired cues. It is not yet known whether this protection from relapse is attributable to antagonism of endogenous cannabinoid tone or to inverse agonism at the CB_1 receptor. The relationship between cannabinoid reward and opiate reward is clear, and some recent work suggests that cannabis use may be useful during treatment for opiate addiction. Finally, the effects of cannabis on the rewarding properties of sexual behavior may differ between men and women; low doses of THC may have positive effects on sexual desire and function in females, but may hinder erectile functioning in men.

7

Cannabinoids, Body Weight, Feeding, and Appetite

One of the most famous effects of marijuana use is the stimulation of appetite, particularly for palatable foods. After the discovery that THC was the psychoactive compound in marijuana, Hollister (1971) verified that a single oral dose of marijuana (containing 0.35 mg/kg THC) increased the intake of milkshakes in healthy volunteers. Subsequently it was demonstrated that smoked marijuana (containing 1.8 percent THC) produced a marked increase in palatable food intake in humans (Foltin, Brady, and Fischman 1986; Foltin, Fischman, and Byrne 1988). Considerable pre-clinical experimental work with rodents demonstrated that the THC and other CB_1-receptor agonists act to enhance appetite both by amplifying the rewarding value of food and by reducing the satiety signals (e.g., leptin secretion) that regulate appetite (Cristino and Di Marzo 2015).

The therapeutic effect of cannabis to stimulate appetite has been studied for decades in the treatment of cachexia, a chronic wasting disorder associated with loss of adipose tissue and lean body mass seen in patients with cancer, patients with acquired immunodeficiency syndrome (AIDS), and patients with anorexia nervosa. Synthetic THC (dronabinol) is used in the clinic to combat the reduction in appetite in such patients, with mixed results. Some studies have found that dronabinol (2.5 mg twice a day) enhanced appetite and body weight in AIDS patients suffering from anorexia (Beal et al. 1997); other studies with both AIDS patients and patients with cancer-related cachexia report no effect with a similar dose and regime (Cannabis-In-Cachexia-Study-Group et al. 2006). Orally administered pure THC may not be the optimal treatment for these disorders. One alternative is the use of FAAH inhibitors (Fegley et al. 2005), which have been shown to enhance motivation for food and to promote energy storage in pre-clinical animal studies (Tourino, Oveisi, Lockney, Piomelli, and Maldonado 2010). As we learn more about the effects of cannabinoids and direct manipulations of the endocannabinoid system on appetite, better treatments may be developed.

Endocannabinoids and Regulation of Body Weight

The potential of the endocannabinoid system to maintain energy balance has been one of its most thoroughly studied effects since the synthetic CB_1 antagonist/inverse agonist SR141716A (rimonabant, also known as Acomplia) was developed by Sanofi-Aventis to treat diet-induced obesity (Arnone et al. 1997). Unfortunately, unwanted psychiatric side effects of depression and suicidal ideation have prevented the use of rimonabant as an anti-obesity treatment. Rimonabant was shown to decrease food intake and body weight gain not only in animal models but also in humans. Pre-clinical work (see, e.g., Koch 2001) demonstrated that stimulation of CB_1 receptors by THC, by synthetic CB_1 agonists, or by endogenous cannabinoids stimulated eating in rats and mice. Genetically modified mice lacking CB_1 receptors did not show this effect. Therefore, the feeding stimulation by THC was dependent on an intact CB_1 system.

In the brain, the endocannabinoid system interacts with the mesolimbic dopamine system to engage with reward pathways and with the hypothalamus to regulate levels of hormones that mediate the enhancement of feeding. Indeed, endocannabinoid levels in the hypothalamus and in reward pathways are highest during food deprivation, leading to food-seeking behaviors.

Endocannabinoid regulation of metabolic feeding-related hormones

Hormonal and nutritional signals in the hypothalamus inform the brain about the free and stored levels of fuel available for the organism. The neural circuitry of the hypothalamus uses this information to regulate caloric intake, energy consumption, and peripheral lipid and glucose metabolism. The actions of endocannabinoids in the hypothalamus serve to quickly fine-tune energy intake. Administration of anandamide (AEA) into the hypothalamus induces eating, and endocannabinoid levels in the hypothalamus vary as a function of nutritional status. The action of endocannabinoids in the hypothalamus regulates the levels of several biochemical compounds that control feeding and body weight, including melanocortin (which reduces feeding) and neuropeptide Y (which stimulates feeding). Within the hypothalamus, modulation of the expression of several pro-feeding and anti-feeding hormones by the endocannabinoid system is counterbalanced by the opposite actions mediated by the adipose-derived hormone leptin (Cristino and Di Marzo 2015).

Leptin originates in adipose tissue and affects a number of appetite-related factors in the hypothalamus; it plays an important role in regulating appetite, hunger, and metabolism. Endocannabinoid levels in the hypothalamus correlate inversely with leptin plasma levels (Di Marzo et al. 2001). Thus, leptin administration (which exerts an anorectic action) suppresses hypothalamic endocannabinoid levels in healthy animals, but in the hypothalamus of obese and hyperphagic rodents lacking leptin—such as ob/ob (obese/obese, leptin mutant) mice (see figure 7.1) and Zucker[(fa/fa)] (leptin-receptor mutant) rats—endocannabinoid levels are significantly increased (ibid.). Indeed, recent evidence indicates that functional CB_1-receptor signaling in the hypothalamus is required for leptin to exert its suppressive effects on food intake; selective genetic knockout of CB_1 receptors in the hypothalamus of mice abolished the inhibition of food intake by leptin (Cardinal et al. 2012). This inverse relationship also affects the activity of the brain's reward system; that is, obese rats with defective leptin signaling show increased CB_1 expression and binding activity in the brain's reward structures (D'Addario et al. 2014).

Besides the hypothalamus and reward system, the endocannabinoid system regulates food intake and energy balance in the vagus nerve, which carries information between the digestive system of the gut and the brainstem. Within the gut, cholesystokinin (CCK) acts to suppress feeding (a satiety signal) and ghrelin acts to stimulate feeding.

Figure 7.1
An ob/ob (leptin-receptor mutant) mouse and a mouse with normal body weight.

CB$_1$-receptor expression in vagal afferents is increased in fasting and decreased after refeeding, under the control of CCK (Burdyga et al. 2004). The decrease in CB$_1$-receptor expression after refeeding is prevented by administration of ghrelin. Thus ghrelin opposes the action of CCK on CB$_1$-receptor expression on vagal afferents to the brainstem. In addition, the CB$_1$-receptor antagonist rimonabant abolishes ghrelin-induced feeding by decreasing levels of circulating ghrelin. Ghrelin has also been shown to elevate hypothalamic endocannabinoid content (D'Addario et al. 2014).

Endocannabinoids and the brain's appetite reward mechanisms

Only one clinical study has investigated the regulation of endocannabinoid levels in humans eating food for pleasure (Monteleone et al. 2012). In normal-weight satiated healthy participants, plasma levels of 2-AG and ghrelin, but not AEA, OEA, and PEA, were elevated after hedonic eating. In addition, 2-AG levels in the hypothalamus increased in mice showing a preference for a high-fat diet (Higuchi et al. 2011). Highly palatable foods may evoke alterations in the central nervous system similar to those evoked by drugs of abuse through the regulation of common neurobiological substrates. Indeed, overconsumption of palatable foods is accompanied by the stimulation of the brain's dopaminergic and opioid reward systems (D'Addario et al. 2014).

As was noted and illustrated in figure 6.3, the mesolimbic dopamine system is the brain's primary reward system. All drugs of abuse activate this system (Wise 2004). The regulation of reward-related appetite processes may be controlled, in part, by endocannabinoid release in the VTA of the midbrain, which produces inhibition of the release of GABA, thus removing the inhibitory effect of GABA on dopaminergic neurons that project to the NAc (Maldonado et al. 2006). In the NAc, released endocannabinoids act on CB$_1$ receptors on axon terminals of glutamatergic neurons. The resulting reduction in the release of glutamate on GABA neurons projecting to the VTA results in indirect activation of the VTA's dopamine neurons. Microinjections of THC into the VTA and the NAc serve as rewards for both self-administration and conditioned place preference in rats (Zangen, Solinas, Ikemoto, Goldberg, and Wise 2006).

In pre-clinical rodent studies, the CB$_1$ inverse agonist/antagonist rimonabant reduces the rewarding effects of foods and drugs. Rimonabant leads to a reduction in a conditioned place preference for palatable food (Chaperon, Soubrie, Puech, and Thiebot 1998). It also reduces the motivation for self-administration of beer in rats (Gallate, Saharov, Mallet, and McGregor 1999). CB_1 antagonism also counteracts the increase of extracellular DA release induced by highly palatable foods, whereas CB_1 agonists do the opposite (Higgs, Barber, Cooper, and Terry 2005; Melis and Pistis 2007). THC increases hedonic reactivity to sucrose (Jarrett, Limebeer, and Parker 2005), and CB_1 antagonism does the opposite (Jarrett, Scantlebury, and Parker 2007). In addition, intra-NAc injections of AEA enhance sucrose hedonics (Mahler, Smith, and Berridge 2007). Thus CB_1 is an important component of the neural substrate that mediates the reinforcing and motivational properties of a highly palatable food. Indeed, CB_1 mRNA expression is downregulated in areas of the limbic forebrain of rats fed a palatable diet (Harrold, Elliott, King, Widdowson, and Williams 2002). This reduction in CB_1 expression represents a compensatory mechanism aimed at counteracting increased levels of endocannabinoids resulting from the consumption of fat-rich palatable food. On the other hand, fasting leads to an increase of limbic AEA and 2-AG levels, which return to normal after re-feeding. This effect occurs only in brain areas not involved in the regulation of feeding behavior (Kirkham, Williams, Fezza, and Di Marzo 2002). This over-activation of the endocannabinoid systems during fasting may be part of a physiological mechanism aimed at enhancing the motivation for food.

Endocannabinoids also have important functional relationships with the endogenous opioid system, which also mediates the rewarding value of food (Kirkham 1991). Interestingly, the facilitory effect of THC on food intake is not only reduced by a CB_1 antagonist; it is also reduced by the µ-opioid antagonist naloxone (Gallate and McGregor 1999). Indeed, a synergistic effect of very low doses of drugs that block the CB_1 receptor and the µ-opioid receptor (Kirkham and Williams 2001) may serve as a promising treatment for obesity.

The Endocannabinoid System, Eating Disorders, and Obesity

Because the endocannabinoid system is a modulator of both homeostatic and hedonic aspects of eating, dysfunctions of that system may lead to eating disorders. Anorexia nervosa and bulimia nervosa are characterized by abnormal eating behaviors resulting in severe food restriction, anorexia nervosa with a dramatic loss of body weight and bulimia nervosa with episodes of binge eating and vomiting without significant changes in body weight. The fifth edition of the *Diagnostic and Statistical Manual of Mental Disorders* includes binge-eating disorder, which is characterized by binge eating but without compensatory vomiting resulting in obesity. Such eating disorders clearly involve both psychosocial and biological factors.

On the basis of the known role of the endocannabinoids in regulation of feeding and energy homeostasis, it is conceivable that dysfunctions in this regulatory system may be involved in the pathophysiology of eating disorders. Monteleone et al. (2005) compared the blood levels of AEA and 2-AG of women with one of these eating disorders with those of healthy controls. They also examined the relationship between the endogenous cannabinoids and circulating leptin (a peripheral fat hormone involved in long-term modulation of body weight and energy balance), nutritional variables, and psychopathological variables. Increased blood levels of AEA (but not of 2-AG) were found in patients with anorexia nervosa or binge-eating disorder, but not in women affected with bulimia nervosa. In addition, anorexic women also showed decreased circulating leptin levels, and binge-eating disorder women showed increased leptin levels relative to healthy controls. In both healthy controls and women with anorexia nervosa, the higher the blood AEA levels the lower were the plasma leptin concentrations. This suggested that the decreased leptin signaling of underweight anorexia nervosa patients could be involved in the increase of AEA levels (Monteleone et al. 2005). Higher blood levels of CB_1 mRNA are detected in women suffering from both anorexia nervosa and bulimia nervosa (Frieling et al. 2009), but decreased peripheral CB_1 mRNA is found in patients with more severe forms of the disorders. More recently, PET scans demonstrated increased CB_1-receptor levels in the insula (a cortical region regulating feeding) and in the inferior frontal and temporal cortex of underweight anorexia nervosa patients and in the insula of women with bulimia nervosa (Gerard, Pieters, Goffin, Bormans, and Van Laere 2011). It is possible that the dysregulated endocannabinoid tone of anorexia nervosa and bulimia nervosa patients may represent an adaptive response aimed at maintaining energy balance by potentiating internal food-seeking signals and hence stimulating food ingestion (Monteleone and Maj 2013). However, most human studies using dronabinol (e.g., Andries, Frystyk, Flyvbjerg, and Stoving 2014; Gross et al. 1983) have reported no improvement of anorexia symptoms.

There appears to be a positive association between obesity and either overproduction of endocannabinoids or increased expression of the CB_1-receptor tissues (central and peripheral) involved in energy homeostasis in both animal and human studies. Genetically obese (ob/ob) mice (figure

7.1) were found to have hypothalamic endocannabinoid levels higher than in lean animals. These effects were significantly reduced by the exogenous administration of leptin (Di Marzo et al. 2001). Subsequently, an elevated CB_1 expression was observed in the white adipose tissue of mice defective in leptin signaling relative to lean controls. Hyperglycemia and insulinemia may cause an over-activation of the endocannabinoid system in obesity-related type-2 diabetes. Dysregulation of the endocannabinoid system has also been reported in overweight obese women with a binge-eating disorder and in obese postmenopausal women. Obesity-related elevations of endocannabinoids were coupled to decreased activity of FAAH, and consistently specific polymorphisms of the FAAH gene were associated with overweight and obesity or with lower insulin sensitivity. Higher levels of 2-AG in adipose tissue were found in samples from patients with elevated abdominal fat distribution relative to patients with subcutaneous fat or lean controls. In addition, AEA levels are elevated in the saliva of obese subjects and directly correlate with Body Mass Index, waist circumference, and fasting insulin (Matias et al. 2012). In these participants, body-weight loss after a 12-week program decreased salivary AEA levels. Therefore, salivary AEA may be a useful biomarker for obesity.

Manipulation of the Endocannabinoid System for Treatment of Obesity

Since there appears to be pathological over-activation of the endocannabinoid system in overweight and obese subjects, current therapeutic targets for obesity are aimed at the restoration of a normal endocannabinoid tone by means of drugs that interfere with endocannabinoid signaling.

The rimonabant story

Initial pre-clinical studies conducted with rimonabant support the hypothesis that antagonism of the endocannabinoid system with the CB_1 inverse agonist/ antagonist rimonabant may be a promising therapy for obesity. Several clinical trials followed the initial experimental findings from the four Rimonabant in Obesity (RIO) studies (Despres, Golay, and Sjöström 2005; Pi-Sunyer et al. 2006; Scheen et al. 2006; Van Gaal et al. 2005). Rimonabant was marketed as an anti-obesity agent in several European countries. In the United States, the Food and Drug Administration asked for further evidence regarding safety before approving its marketing. A subsequent meta-analysis of the four RIO studies suggested that patients treated with rimonabant were 2.5–3 times as likely to experience psychiatric adverse effects, such as anxiety and depression, as patients receiving placebo (Christensen, Kristensen, Bartels, Bliddal, and Astrup 2007). The European Union Committee for Medicinal Products for Human Use concluded that rimonabant doubled the risk for psychiatric disorders, and the European Medicines Agency (EMA) suspended the license for the drug. Sanofi-Aventis withdrew rimonabant from the worldwide market, and clinical development not only of rimonabant but also of other CB_1 antagonists being developed by other companies was halted.

New potential treatments for obesity exploiting the eCB system

Recent pre-clinical investigation has focused on the evaluation of CB_1 antagonists/inverse agonists that do not penetrate the brain as potential anti-obesity treatments. Such treatments may be devoid of the adverse psychiatric side effects produced by brain-penetrate CB_1 antagonists/ inverse agonists, such as rimonabant. Human adipose tissue has been shown to possess a fully functional endocannabinoid system (Spoto et al. 2006), and a high-fat diet induces an increase of AEA in the liver as a result of reduced degradation by FAAH (Osei-Hyiaman et al. 2005). In addition, 2-AG levels are elevated in the visceral fat of obese patients (Bluher et al. 2006). These findings suggest that the hypophagic effect of rimonabant may be mediated by its action on peripheral CB_1 receptors rather than central receptors. (See also Gomez et al. 2002.) Therefore, the beneficial effects of blocking peripheral CB_1 receptors to reduce appetite, body weight, hepatic steosis, and insulin resistance are being investigated (Tam et al. 2010; Tam et al. 2012). Tam et al. (2010) found that the CB_1- receptor antagonist (without inverse agonist properties) AM6545 was as effective as rimonabant in ameliorating the fatty liver and improving the lipid profile in mice with diet-induced obesity, but that it was less effective than rimonabant in reducing body weight, adiposity, insulin resistance, and hyperleptinemia and had minimal effect on food intake. Because rimonabant is both a brain penetrant and a CB_1-receptor inverse agonist, its greater efficacy could be due either to its central action or to its inverse agonist properties. Tam et al. (2012) demonstrated that it is the latter property that makes rimonabant a better treatment for obesity and its related metabolic disorders. That is, with the use of a peripherally restricted CB_1-receptor inverse agonist, JD5037, they demonstrated as much reduction in food intake, body weight, and adiposity as was obtained with rimonabant. These results are particularly exciting because they may lead to new treatments for obesity without the possibility of psychiatric side effects that prevented the use of rimonabant as a treatment.

Another approach to reducing the impact of endocannabinoids on feeding is to interfere with their metabolism. Indeed, decreasing 2-AG levels by systemic administration of DAGL inhibitor (the enzyme responsible for the synthesis of 2-AG) reduced intake of palatable or high-fat foods by mice (Bisogno et al. 2009; Bisogno et al. 2013). In addition, mice genetically altered to overexpress MAGL (the enzyme that degrades 2-AG) and put on a high-fat diet showed increased energy expenditure and decreased weight gain (Jung et al. 2012). Another strategy that is being explored is related to the allosteric modulation of the CB_1 receptor by changing the binding of the endocannabinoid. PSNCBAM-1, a novel allosteric antagonist of the CB_1 receptor, reduces that impact of AEA or that of 2-AG when it binds to the orthosteric receptor site. This allosteric antagonist inhibited appetite and produced weight loss in rats (Horswill et al. 2007). Furthermore, there is some evidence that CB_2 antagonism has potential to manage obesity-associated metabolic disorders (Deveaux et al. 2009); this is important because CB_2 agonists do not produce central effects.

What About the Other Cannabinoids in Cannabis?

THC clearly has the potential to enhance appetite through its agonism of the CB_1 receptor. However, its psychoactive side effects limit its usefulness as an appetite-stimulating agent in treating anorexia or cachexia. What about the other cannabinoids found in cannabis that are not psychoactive?

THCV
In view of the withdrawal from the market of rimonabant, taranabant, and other CB_1 inverse agonists because of unwanted psychiatric side effects, safer alternatives are needed. THCV is a CB_1 antagonist without inverse agonist effects that reduces food intake in rats (Pertwee 2007; Pertwee 2008; Thomas et al. 2005). The lack of inverse agonism of the CB_1 receptor with THCV may render it more tolerable, because preclinical research suggests that the psychiatric side effects of rimonabant were produced by its central inverse agonism (not antagonism) of the CB_1 receptor (Limebeer et al. 2010; Sink et al. 2008). In addition, CB_1 antagonists, unlike inverse agonists, are devoid of potential anhedonic effects

(Meye, Trezza, Vanderschuren, Ramakers, and Adan 2013) and nausea-producing effects (Limebeer et al. 2010; Sink et al. 2008). Therefore, it will be interesting to see whether THCV will be developed as a potential appetite-reducing anti-obesity drug.

CBD

Early work with CBD indicated that at a high dose (50 mg/kg, ip) it reduced consumption in rats trained to consume their daily food intake in 6 hours; when lab chow was replaced with sucrose, however, suppression of intake was greatly decreased (Scopinho, Guimaraes, Correa, and Resstel 2011; Sofia and Knobloch 1976; Wiley et al. 2005). More recent work has indicated that CBD (3–100 mg/kg, ip) does not modify food intake in mice (Wiley et al. 2005; Scopinho et al. 2011). However, when administered orally, relatively low doses of CBD have been shown to produce a short-term reduction in feeding (Farrimond, Whalley, and Williams 2012) by an unknown mechanism. Since CBD has been shown to ameliorate some unwanted effects of THC (e.g., anxiety and learning deficits), it may be worthwhile to evaluate combined doses of CBD and THC in treatment of anorexia nervosa or other eating disorders.

Conclusion

At the turn of the twenty-first century, new medications based on inverse agonism/antagonism of the CB_1 receptor showed great promise in treating obesity. However, shortly after such medications were widely prescribed in several countries, patients reported suicidal thoughts and anxiety, probably attributable to inverse agonism of central CB_1 receptors.

It is clear that the endocannabinoid system is an important regulator of appetite, food preference, and body weight. The endocannabinoid system not only regulates metabolic feeding-related hormones in the brain and in the gut, but also regulates the brain reward circuitry involved in palatability-based feeding. One of the primary roles of the endocannabinoid system is in the homeostatic regulation of feeding behavior.

New treatments for obesity being developed by chemists and pharmaceutical companies attempt to harvest the anti-obesity effects of rimonabant, but are devoid of the psychiatric side effects that became clearly known only after rimonabant was widely prescribed.

8

Cannabinoids and Nausea

Nausea and vomiting are important to ensure survival in hostile external environments that often contain contaminated or potentially harmful foods. Nausea and vomiting also are easily activated in various disease states, such as diabetes and labyrinthitis (Sharkey et al. 2014). Migraine headaches, concussions, and other conditions of the central nervous system also can cause nausea and vomiting. Nausea and vomiting are often side effects of medications used to treat many conditions, most notably of agents used in chemotherapy for cancer.

Nausea and vomiting often occur shortly after chemotherapy treatment. Current anti-emetic therapies are highly effective in reducing chemotherapy-induced vomiting, but they are only weakly effective in treating chemotherapy-induced nausea. Such nausea can be so severe that as many as 20 percent of patients discontinue chemotherapy, even when vomiting is pharmacologically controlled (Andrews and Horn 2006). Cancer patients receiving chemotherapy treatment also often experience delayed (24–48 hours) nausea and vomiting, which are not well controlled by available anti-emetic treatments. When the initial acute emetic episode is not prevented, patients often experience anticipatory nausea and vomiting upon returning to the clinic. At present, the only available treatments for anticipatory nausea and/or vomiting are non-selective sedatives

Although the mechanisms of vomiting are well known (Hornby 2001), those of nausea remain poorly understood. For that reason, effective treatments are very limited. Yet nausea is reported to be one of the most debilitating human sensations, and thus new treatments are urgently needed.

THC and Chemotherapy-Induced Nausea and Vomiting in Humans

The treatment of nausea and vomiting was the first recognized medical use of THC in modern medical history. Ineffective treatment of chemotherapy-induced nausea and vomiting prompted oncologists to investigate the anti-emetic properties of cannabinoids in the late 1970s and the early 1980s, before the discovery of antagonists of 5-hydroxytryptamine-3 (5-HT$_3$) receptors such as ondansetron (Costall, Domeney, Naylor, and Tattersall 1986; Miner and Sanger 1986). In 1981 the cannabinoid agonist nabilone (Cesamet), a synthetic analogue of THC, was specifically licensed for the suppression of nausea and vomiting produced by chemotherapy treatment. Synthetic THC (dronabinol) entered the clinic as Marinol in 1985 as an anti-emetic and in 1992 as an appetite stimulant (Pertwee 2009). In the early studies, several clinical trials compared the effectiveness of THC with placebo or with other anti-emetic drugs that were available at the time, most commonly dopamine antagonists. Nabilone produced fewer vomiting episodes and reports of nausea than metoclopramide (D2 antagonist) in patients with moderately toxic chemotherapy treatments. However, as the severity of the nausea and vomiting increased with highly emetogenic treatments (cisplatinum), nabilone and metoclopramide showed equivalent efficacy in reducing nausea and vomiting. Marinol is prescribed as an anti-emetic agent and as an appetite stimulant. When some patients were given the D2-receptor antagonist prochlorperazine alone and others were given both prochlorperazine and dronabinol, patients receiving the combined treatment had less severe chemotherapy-induced nausea (with a moderate emetic treatment) than those treated with either drug alone. (For reviews, see Rock, Limebeer, Sticht, and Parker 2014 and Tramer et al. 2001.)

There is also evidence that cannabis-based medicines may be particularly effective in treating the harder-to-control symptoms of nausea and delayed nausea and vomiting in children. Abrahamov et al. (1995) evaluated the anti-emetic effectiveness of Δ^8-THC, a close but less psychoactive relative of Δ^9-THC, in children receiving chemotherapy treatment. Two hours before the start of each cancer treatment and every six hours thereafter for 24 hours, the children were given Δ^8-THC as oil drops on the tongue or in a bite of food. After 480 treatments, the only side effects reported were slight irritability in two of the youngest children (one 3½ years old, the other 4). Yet both acute and delayed nausea were controlled very well (Abrahamov, Abrahamov, and Mechoulam 1995).

This small-scale study (without a placebo control) provides suggestive evidence that Δ^8-THC may be particularly effective in alleviating chemotherapy-induced nausea and vomiting in children with cancer. It is surprising that there have been no follow-up clinical trials to the single study.

A major advance in the control of acute vomiting in chemotherapy treatment came in the late 1980s with the development of antagonists of 5-hydroxytryptamine–3 (5-HT$_3$) receptors, such as ondansetron. These agents were highly effective in suppressing vomiting in pre-clinical studies using ferrets and shrews. (Rats and mice are incapable of vomiting.) In clinical trials with humans, treatment with 5-HT$_3$ antagonists (often in combination with the corticosteroid dexamethasone) during the first chemotherapy treatment reduced the incidence of acute vomiting by approximately 70 percent (Parker, Rock, and Limebeer 2011; Sharkey et al. 2014). However, as the use of these 5-HT$_3$ antagonists in the clinic progressed it became clear that they are much more effective in reducing acute vomiting than in suppressing acute nausea (Hickok et al. 2003). These compounds are also ineffective for treating delayed nausea and vomiting that occurs 24–48 hours after the treatment and for treating anticipatory nausea experienced as a conditioned response when the patient returns to the clinic in which the treatment occurred (Morrow and Dobkin 1987). More recently, NK$_1$-receptor antagonists (e.g., Aprepitant) have been developed that not only decrease acute vomiting but also decrease delayed vomiting induced by cisplatin-based chemotherapy (Van Belle et al. 2002); however, these compounds alone and in combination with 5-HT$_3$ antagonist/dexamethasone treatment are also much less effective in reducing nausea (Hickok et al. 2003), the symptom reported to be the most distressing to patients undergoing treatment with 5-HT$_3$ antagonists (de Boer-Dennert 1997). Considerable evidence suggests that another system that may be an effective target for treatment of chemotherapy-induced nausea, of delayed nausea/vomiting, and of anticipatory nausea/vomiting is the endocannabinoid system. (For a review, see Sticht, Rock, Limebeer, and Parker 2015.)

Surprisingly, only one clinical trial (Meiri et al. 2007) has compared the anti-emetic and anti-nausea effects of cannabinoids with those of the more recently developed 5-HT$_3$ antagonists, and none has compared cannabinoids with the NK$_1$ antagonist aprepitant. Meiri et al. (ibid.) compared the efficacy and tolerability of dronabinol with ondansetron or the combination for delayed chemotherapy-induced nausea and vomiting in a five-day, double-blind, placebo-controlled trial. Patients receiving moderately to highly emetogenic chemotherapy were all given both ondansetron and dexamethasone; in addition, half of them received placebo and the other half received dronabinol before chemotherapy on day 1. On days 2–5, some received placebo, some dronabinol, some ondansetron, and some both dronabinol and ondansetron. The results revealed that the efficacy of dronabinol alone was comparable with that of ondansetron in the treatment of delayed nausea and vomiting for the total response of no vomiting and nausea; however, the dronabinol group reported the lowest nausea intensity on a visual analog scale. The dose of dronabinol used in this study was at least 50 percent lower than in previous studies conducted in the 1980s, resulting in a low incidence of CNS-related adverse effects, which did not differ from the incidence in the ondansetron-treated group. The results of this study suggest that THC may be more effective than 5-HT_3 antagonists in treating delayed nausea.

Recently an oromucosal cannabis-based medicine, Sativex (2.7 mg THC: 2.5 mg cannabidiol/spray) has been evaluated in phase II clinical trials for its potential to enhance the effectiveness of standard anti-emetic treatment for chemotherapy-induced nausea and vomiting (Duran et al. 2010). Sativex or placebo was administered to patients also receiving a 5-HT_3 antagonist and a corticosteroid. Duran and colleagues found that Sativex facilitated relief of delayed nausea and vomiting among patients receiving moderately emetogenic cancer chemotherapy. In particular, 57 percent of the Sativex-treated patients experienced no delayed nausea and 71 percent experienced no delayed emesis. However, in view of the known anti-nausea and anti-emetic effects of CBD demonstrated in animal models (Parker, Mechoulam, and Schlievert 2002; Parker, Rock, Sticht, Wills, and Limebeer 2015; Rock, Limebeer, Mechoulam, Piomelli, and Parker 2008), it is not clear to what extent the anti-nausea/anti-emetic effects of Sativex are attributable to either THC or CBD; nonetheless, this study demonstrates the therapeutic potential of combining these cannabinoids in treating delayed nausea and vomiting.

Most of the reported clinical trials of the effectiveness of cannabinoid compounds in treating chemotherapy-induced nausea and vomiting have involved orally delivered cannabinoids, which may be less effective than inhaled cannabinoids. Many patients have a strong preference for smoked marijuana over orally delivered synthetic cannabinoids (Tramer et al. 2001). Several reasons for this have been suggested, including the advantages of self-titration of the dose, difficulty in swallowing the pills while experiencing emesis, faster onset with inhaling than with oral delivery, and the combined action of other cannabinoids (especially CBD) that are found in marijuana. Although many marijuana users have claimed that smoked marijuana is a more effective anti-emetic than oral THC, no controlled studies have yet been published that specifically evaluate this possibility.

Despite the evidence for the anti-nausea potential of cannabinoids in humans, it is worthwhile to note that chronic marijuana use has also been reported to produce the sensation of nausea, and also to produce vomiting. This paradoxical effect of marijuana, known as *cannabinoid hyperemesis syndrome* (Sullivan 2010) has been documented in numerous case reports in recent years (Allen, de Moore, Heddle, and Twartz 2004; Roca-Pallin, Lopez-Pelayo, Sugranyes, and Balcells-Olivero 2013; Simonetto, Oxentenko, Herman, and Szostek 2012). Although the cause of hyperemesis is not known, changes in CB_1-receptor expression and variations in concentrations of THC (and other marijuana constituents) may figure in this paradoxical effect. Further research is needed to fully understand the mechanism(s) underlying this peculiar effect of cannabinoids, which may lead to increases in nausea and in vomiting after chronic marijuana consumption. Nevertheless, these cases highlight the dynamic nature of the role of the endocannabinoid system in regulating nausea and suggest a possible consequence of endocannabinoid-system dysregulation, which may lead to undesirable results such as an increased sensation of nausea and emesis.

Phytocannabinoids and Vomiting and Nausea in Pre-Clinical Animal Models

Considerable experimental pre-clinical animal evidence accumulated since the discovery of the mechanism of action of THC indicates that THC not only reduces vomiting but also reduces both acute and anticipatory nausea, two symptoms of chemotherapy treatment that are more difficult to treat than vomiting. And promising new research suggests that CBD and other components of cannabis may also reduce acute and anticipatory nausea.

THC and vomiting and nausea in animal models

THC reduces toxin-induced vomiting in cats, dogs, ferrets (Van Sickle et al. 2001), and shrews (Darmani 2001a). Besides reducing acute vomiting produced by cisplatin, THC attenuates delayed vomiting in shrews (Ray, Griggs, and Darmani 2009). The effects of THC on vomiting are consistently reversed by CB$_1$ antagonists, which verifies a CB$_1$-mediated mechanism of action (Parker et al. 2011; Sharkey et al. 2014). Indeed, the CB$_1$ inverse agonist/antagonist rimonabant produces vomiting in shrews, which is a function of its inverse agonist effects (Darmani 2001b). Interestingly, when sub-threshold doses of THC and ondansetron are combined, they produce a synergistic effect that completely suppresses cisplatin-induced vomiting in shrews (Kwiatkowska, Parker, Burton, and Mechoulam 2004). The anti-emetic properties of CB$_1$ agonists are mediated by their action on CB$_1$ receptors in the emetic brainstem structures located in the dorsal vagal complex (DVC) (Parker 2014; Van Sickle et al. 2001). Therefore, both the pre-clinical evidence and the human evidence indicate an anti-emetic role for CB$_1$ agonists.

Nausea is much more difficult to control with the currently available anti-emetic medications than is vomiting. One reason for our lack of understanding of nausea is that until recently there were no validated animal models for pre-clinical evaluation of new treatments for nausea. Recent evidence based on the development of new rat models of nausea (Parker 2014) suggests that cannabinoids may have therapeutic potential for reducing both acute and anticipatory nausea. Although rats do not vomit, their brains receive the same physiological signals produced by emetic toxins in the gut and in the brainstem emetic regions as ferrets. Grill and Norgren (1978, 1981) identified a unique behavioral marker of nausea in rats that they called *conditioned gaping*. Conditioned gaping, defined as the wide opening of the mouth exposing the lower incisors (see figure 8.1), requires the same orofacial musculature as vomiting in

Figure 8.1
Left: rat gape. Right: orofacial component of a shrew's retch just before it vomits.

an emetic species and is topographically similar to the orofacial components of retching in the shrew (Travers and Norgren 1986). Of course,
no one can know if a rat experiences the same symptoms as a human
who subjectively reports a sensation of nausea. However, considerable
behavioral evidence confirms that only manipulations that produce vomiting in other species promote conditioned gaping in rats and that anti-
nausea drugs consistently prevent the establishment of conditioned
gaping in rats (Parker 2014). Typically, conditioned gaping in rats is produced by pairing a flavor with an emetic treatment, but recently it has
been shown that contextual stimuli can also elicit conditioned gaping
when paired with an emetic treatment. (See Rock, Limebeer, and Parker
2014 for a review.) This contextually elicited conditioned gaping in rats
provides a model of anticipatory nausea seen in humans when they
return to the clinic in which they received their chemotherapy treatment.
In rats, as is seen in human cancer patients experiencing anticipatory
nausea, contextually elicited conditioned gaping is not suppressed with
conventional anti-emetic treatments, such as the 5-HT_3 antagonist
ondansetron; however, cannabinoid treatments are consistently effective
in this model. These pre-clinical rat models of acute and contextually
elicited anticipatory nausea have provided considerable evidence for the
potential of cannabinoid treatments to reduce both acute and anticipatory nausea.

THC and other CB_1 agonists consistently interfere with acute nausea
in the rat gaping model by acting at CB_1 receptors (Rock, Limebeer, and
Sticht 2014). The CB_1 inverse agonist/antagonist rimonabant produces
nausea in the rat gaping model (McLaughlin et al. 2005), but CB_1 antagonists without inverse agonist properties (e.g., AM4113) do not produce
nausea in that model (Sink et al. 2008). In fact, rimonabant has been
shown to enhance the nauseating effects of toxins (Parker et al. 2003).
The site of the anti-nausea effect of CB_1 agonists appears to be the interoceptive insular cortex, which receives afferents from the emetic circuitry
of the brain (Limebeer, Rock, Mechoulam, and Parker 2012; Sticht,
Limebeer, Rafla, Abdullah, et al. 2015).

Not only do rats display conditioned gaping reactions when re-exposed
to a flavor previously paired with a nausea-inducing drug; they also display conditioned gaping reactions when re-exposed to a context previously paired with a nausea-inducing drug. (See Rock, Limebeer, and
Parker 2014 for a review.) Contextually elicited conditioned gaping in
rats is remarkably similar to anticipatory nausea experienced by chemotherapy patients when they return to the context in which they received

their chemotherapy. When anticipatory nausea develops in human chemotherapy patients, the classic anti-emetic drug ondansetron is ineffective in reducing the symptom (Hickok et al. 2003); likewise, rats treated with ondansetron do not show suppression of contextually elicited conditioned gaping. On the other hand, THC and other CB_1-receptor agonists effectively suppress anticipatory nausea in the rat anticipatory nausea model. (For reviews, see Rock, Limebeer, and Parker 2014 and Rock, Limebeer, Sticht, et al. 2014.)

CBD and nausea and vomiting
No study has evaluated the potential of CBD to reduce nausea and vomiting in human chemotherapy patients. However, the pre-clinical literature shows that CBD reduces toxin-induced vomiting in house musk shrews (Kwiatkowska et al. 2004), with a biphasic effect. Low doses (5–10 mg/kg, ip) suppress vomiting, but high doses (40 mg/kg, ip) increase it. In the rat gaping model, CBD is also effective in reducing both acute nausea (Parker et al. 2002) and anticipatory nausea (Rock et al. 2008). The effect of CBD on acute nausea appears to be mediated by its action on 5-HT_{1A} receptors located in the dorsal raphe nucleus (Rock et al. 2012). Action of 5-HT_{1A} agonists at that site reduces the release of forebrain serotonin by acting at somatodendritic autoreceptors. Since elevated forebrain serotonin may be a trigger for nausea, the reduction in its release by CBD may be the mechanism for its anti-nausea effects.

Combined THC and CBD
Sativex taken in conjunction with ondansetron has been evaluated in a phase II clinical trial for its ability to control delayed chemotherapy-induced nausea and vomiting (Duran et al. 2010). In addition, a combination of CBD and THC produces a synergistic effect in suppressing vomiting in shrews and acute nausea in rats; that is, combined sub-threshold doses of CBD and THC completely blocked toxin-induced vomiting and nausea in the animal models (Rock, Limebeer, and Parker 2015).

THCA and CBDA
THCA and CBDA, the non-psychoactive carboxylic precursors of THC and CBD, are present in fresh cannabis plant before it is heated. THCA is ten times as potent as THC in suppressing vomiting in shrews and acute nausea in rats (Rock, Kopstick, et al. 2013), with a CB_1 mechanism of

action. CBDA is 100–1,000 times as potent in reducing vomiting, acute nausea, and anticipatory nausea in these models (Bolognini et al. 2013; Rock and Parker 2013), with a $5\text{-}HT_{1A}$ mechanism of action. Interestingly, CBDA and ondansetron act synergistically in such a way that, when combined, sub-threshold (ineffective) doses of each completely abolish both vomiting and acute nausea in these models. In addition, a synergistic effect is seen with combined sub-threshold doses of CBDA and THC when delivered systemically or orally in both the rat model of acute nausea and the rat model of anticipatory nausea (Rock, Limebeer, and Parker 2015). CBDA alone and in combination with THC or ondansetron is extremely promising as a treatment for acute nausea. CBDA is especially promising as a treatment for anticipatory nausea, a symptom for which the only available treatment is anti-anxiety medication with sedative side effects.

THCV, CBDV, and CBG

THCV is a CB_1 antagonist at lower doses, but can act as a CB_1 agonist at higher doses. The action of THCV is not one of inverse agonism, but one of antagonism (Rock, Sticht, et al. 2013). Unlike rimonabant (an inverse agonist of CB_1), THCV does not produce nausea on its own. In fact, it appears to act as an anti-nausea agent. Unlike rimonabant, which enhances LiCl-induced nausea, THCV reduces nausea at high doses (Rock, Sticht, et al. 2013). Although very little is understood about the actions of CBDV, it appears to reduce nausea at high doses. On the other hand, cannabigerol (CBG) prevented the anti-nausea effects of CBD by acting as a $5\text{-}HT_{1A}$ antagonist (Rock et al. 2011). Thus, various compounds found in the cannabis plant often seem to oppose one another (Pertwee 2009).

Endocannabinoids and Nausea and Vomiting

Systemic administration of cannabinoids (oral, smoking, or sub-lingual) produces global activation of the CB_1 receptors. On the other hand, because endocannabinoids are synthesized on demand, manipulations that target endocannabinoid hydrolysis result in a much more localized increase in levels of 2-AG and AEA than systemic administration of cannabinoid-receptor agonists. Therefore, a localized increase in endocannabinoid levels is less likely to produce unwanted side effects attributable to a global effect and is, thus, preferable for selectively reducing nausea and vomiting.

Endocannabinoids during the experience of nausea in humans

There have been no assessments of the potential of treatments that elevate natural endocannabinoids, such as AEA or 2-AG, to reduce nausea and/or vomiting in humans. However, there is evidence that endocannabinoid levels vary in response to nausea-inducing manipulations. General anesthesia often produces postoperative nausea and vomiting. Schelling et al. (2006) investigated the effects of general anesthesia on plasma AEA levels using the volatile inhalant sevoflurane and the intravenously administered anesthetic propofol. Although AEA levels remained unchanged in patients receiving propofol, sevoflurane resulted in a significant decrease in AEA blood levels from the time of induction until 40 minutes later (ibid.). This pattern of results is consistent with reports that propofol acts as an FAAH inhibitor and thus elevates AEA (Patel et al. 2003), which would counteract the effect of the anesthetic on AEA. Indeed, propofol has been reported to produce less postoperative nausea than sevoflurane (Kumar, Stendall, Mistry, Gurusamy, and Walker 2014), perhaps as a result of enhanced AEA signaling. However, this conclusion must be tempered by the recent report (Jarzimski et al. 2012) that sevoflurane and propofol produced similar decreases in plasma AEA. Motion sickness associated with parabolic flight maneuvers has also been shown to reduce AEA and 2-AG levels, but only among participants experiencing nausea from the treatment (Chouker et al. 2010). The fact that participants not experiencing nausea actually had higher levels of AEA than before the manipulation suggests that AEA protected against experiencing nausea in these participants.

Manipulations of the endocannabinoid system and vomiting and nausea in pre-clinical animal models

There has been considerable pre-clinical evaluation of treatments that boost the endocannabinoid system to reduce nausea and vomiting in animal models. Much like the anti-emetic and anti-nausea properties of synthetic and plant-based cannabinoids, treatments that elevate the natural endocannabinoids AEA and 2-AG also directly suppress acute and anticipatory nausea in rat gaping models (Sticht, Rock, et al. 2015). The effects of the exogenous delivery of these endocannabinoids are short-lived, because AEA is rapidly degraded by FAAH and 2-AG is rapidly degraded by MAGL. The FAAH inhibitors URB597 and PF3845 (which elevate AEA for as long as 24 hours) suppress both acute and anticipatory nausea in the rat gaping models, and the more recently developed MAGL inhibitor MJN110 (which elevates 2-AG for 24 hours) also suppresses acute and anticipatory nausea in a CB$_1$-dependent manner (Cross-Mellor et al. 2007; Parker, Niphakis, et al. 2015; Rock et al. 2008; Rock, Limebeer, Ward, et al. 2015). These compounds also suppress toxin-induced vomiting in the shrew model of emesis (Parker et al. 2009; Parker, Niphakis, et al. 2015). More recently, dual FAAH/MAGL inhibitors (e.g., JZL195 and AM4302) have been shown to be highly effective in suppressing anticipatory nausea in rats (Limebeer et al. 2014; Parker et al. 2016). Since conventional anti-emetic treatments are completely ineffective in treating anticipatory nausea and relatively ineffective in treating acute nausea, there is a great need for effective treatments for this distressing side effect of chemotherapy treatment. The results mentioned above suggest that treatments that boost the endocannabinoid system by inhibiting FAAH or MAGL should be tested in human clinical trials for the treatment of acute and anticipatory nausea.

The Role of the Interoceptive Insular Cortex in Endocannabinoid Regulation of Nausea

Brainstem mechanisms clearly are necessary for the generation of a vomiting reflex in animals capable of vomiting, and CB$_1$-receptor activation within that region reduces vomiting (Sharkey et al. 2014; Van Sickle et al. 2001). However, Grill and Norgren (1978), using sub-collicular decerebrate rats, demonstrated that brainstem regions were not necessary for the generation of nausea in rats. Figure 8.2 presents a hypothetical central neural pathway for the generation of nausea. The regions of the brain that mediate nausea-induced conditioned gaping reactions in rats are likely to be the neural regions responsible for the sensation of nausea. Indeed, lesions of the area postrema, of the parabrachial nucleus, and of the insular cortex, but not lesions of the amygdala, interfere with nausea-induced conditioned gaping in rats (Parker 2014), which suggests that these regions may figure in the generation of nausea. Recent research suggests the insular cortex as the forebrain region that may be ultimately responsible for the sensation of nausea.

There is growing evidence that forebrain serotonin (5-HT) generates nausea. Manipulations that reduce forebrain 5-HT availability reduce nausea-induced gaping in rats. Depletion of forebrain 5-HT by selective 5,7-dihydroxytryptamine (5,7-DHT) lesions of the dorsal raphe nucleus and the median raphe nucleus (which produces most of the central serotonin that projects to higher forebrain regions) completely prevents nausea in the rat gaping model (Limebeer, Parker, and Fletcher 2004). In

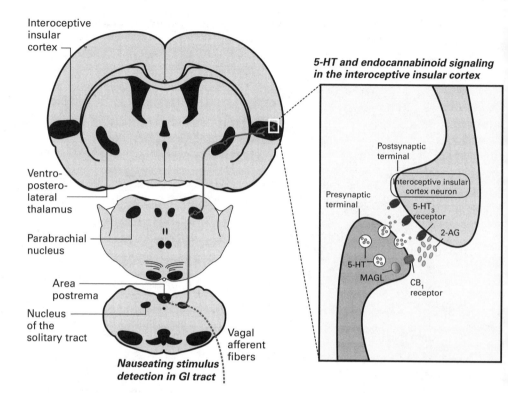

Interoceptive insular cortex

Ventro-postero-lateral thalamus

Parabrachial nucleus

Area postrema

Nucleus of the solitary tract

Vagal afferent fibers

Nauseating stimulus detection in GI tract

5-HT and endocannabinoid signaling in the interoceptive insular cortex

Postsynaptic terminal

Interoceptive insular cortex neuron

Presynaptic terminal

5-HT$_3$ receptor

2-AG

5-HT

MAGL

CB$_1$ receptor

Figure 8.2
Circuits and projections that control nausea. Stimuli that produce emesis activate components in the gut-brain axis to elicit nausea and vomiting (in species capable of the latter). Toxins are detected in the gastrointestinal tract and relay neural input to the brain through vagal afferents terminating in the nucleus of the solitary tract. In addition, blood-borne toxins can gain access to the CNS via the area postrema. Input from the hindbrain is transmitted by afferents to the midbrain parabrachial nucleus by afferents to the ventroposterolateral parvicellular (VPLpc) thalamic nucleus before ultimately reaching the posterior granular layer of the insular cortex (the interoceptive insular cortex). Within the interoceptive insular cortex, we hypothesize that activation of post-synaptic 5-HT$_3$ receptors by serotonin results in the biosynthesis of the endocannabinoid 2-AG, which turns off nausea by reducing the release of nausea-inducing serotonin. When MAGL inhibitors are infused into the interoceptive insular cortex, the action of 2-AG is prolonged for 24 hours, resulting in an effective anti-nausea treatment. (drawing by Martin A. Sticht)

addition, somatodendritic 5-HT_{1A} autoreceptor agonists (8-OH-DPAT, CBD and CBDA), which reduce the rate of firing of afferents that release 5-HT in terminal forebrain regions, were found to suppress conditioned gaping reactions (Bolognini et al. 2013; Limebeer et al. 2004; Rock et al. 2012). Therefore, it is likely that that elevated forebrain serotonin triggers nausea and that reduced forebrain serotonin suppresses it.

Considerable evidence suggests that the insular cortex is the critical region for generating the sensations of nausea and disgust. Electrical stimulation of the insular cortex (IC) produces nausea in humans (Penfield and Faulk 1955). In rats, complete removal of the IC prevents conditioned gaping (Kiefer and Orr 1992). Electrophysiological and anatomical studies have determined that in rats the IC is the cortical site of topographical input of visceral input (posterior granular IC, which is called the interoceptive or visceral IC) and gustatory input (anterior dysgranular IC, which is referred to as the gustatory IC) and their convergence (agranular IC) (Allen, Saper, Hurley, and Cechetto 1991; Cechetto and Saper 1987; Yasui, Breder, Saper, and Cechetto 1991). Although most gastrointestinal visceral sensory afferents to the insular cortex terminate in the interoceptive insular cortex, most efferent projections from the IC to autonomic structures (e.g., the nucleus of the solitary tract in the brainstem) originate in the more anterior agranular field. Inactivation of the interoceptive insular cortex attenuates toxin-induced malaise and toxin-induced Fos activation in this same region, which suggests that it may be responsible for sensing strong deviations from a "well-being state" (that is, normal state) (Contreras, Ceric, and Torrealba 2007). Indeed, recent evidence also implicates the interoceptive insular cortex as a region that mediates addiction as well as nausea (Naqvi, Gaznick, Tranel, and Bechara 2014).

The interoceptive insular cortex is also the site at which serotonin produces nausea-induced conditioned gaping reactions in rats. Tuerke et al. (2012) demonstrated that 76 percent reduction of 5-HT (by 5,7-dihydroxytryptamine lesions) in the entire insular cortex dramatically suppressed nausea in the rat gaping model. Furthermore, they found that intracranial administration of the 5-HT_3-receptor antagonist ondansetron into the interoceptive insular cortex (but not the gustatory insular cortex) attenuated nausea-induced conditioned gaping reactions. The direct delivery of a 5-HT_3-receptor agonist into that region produced the opposite effect (ibid.). These data provide strong evidence that serotonergic input to the interoceptive insular cortex is necessary for the production of acute nausea.

CB$_1$ receptors are also found in the insular cortex. Thus the interoceptive insular cortex is a potential site of the anti-nausea effects of cannabinoids. There is evidence that CB$_1$ receptors are localized on the pre-synaptic terminal endings of serotonin-releasing neurons in other regions of the brain (Hermann, Marsicano, and Lutz 2002). However, it has not been determined if such co-localization occurs in the interoceptive insular cortex, although that seems likely in view of the recent finding that delivery of the potent cannabinoid agonist HU-210 into the interoceptive insular cortex (but not the gustatory insular cortex) reduced LiCl-induced nausea in rats (Limebeer et al. 2012). Furthermore, the aforementioned suppression of nausea-induced behavior was reversed by pre-treatment with the CB$_1$ antagonist AM251, which indicated that it was CB$_1$ mediated. Therefore, cannabinoids may reduce nausea by acting on the CB$_1$ receptors in the interoceptive insular cortex to suppress the nausea-inducing effects of serotonin release in that region.

Recently it has been determined that the endocannabinoid responsible for the regulation of nausea in the interoceptive insular cortex appears to be 2-AG rather than AEA (Sticht, Limebeer, Rafla, Abdullah, et al. 2015; Sticht, Limebeer, Rafla, and Parker 2015). Intracranial administration of 2-AG, but not of AEA, suppressed acute nausea in rats. Furthermore, intra-interoceptive insular cortex infusion of the MAGL inhibitor MJN110, which selectively elevated 2-AG, suppressed toxin-induced nausea, and that effect was reversed by pre-treatment with a CB1 antagonist. Neither the FAAH inhibitor URB597 nor the FAAH inhibitor PF3845 (which elevated AEA in that region) reduced nausea after infusion into the interoceptive insular cortex. Indeed, administration of the nausea-inducing toxin produced a selective elevation of 2-AG, but not of AEA, in the interoceptive insular cortex. Therefore, at least in the interoceptive insular cortex, there is a differential effect of the two ligands for the CB$_1$ receptor in the regulation of nausea: 2-AG, but not AEA, suppressed acute nausea in that region. Strangely, however, elevations in 2-AG produced by MJN110 and elevations in AEA produced by URB597 and by PF3845 all interfered with the expression of anticipatory nausea in this model (Limebeer et al. 2016). Future research is needed to understand the role of central AEA in the regulation of nausea and to understand the difference between its effect on acute nausea and its effect on anticipatory nausea in the interoceptive insular cortex.

Conclusions

The treatment of nausea has lagged behind the treatment of vomiting. There is an immediate and pressing need for new approaches to treat this distressing and debilitating symptom. The development of pre-clinical animal models of acute and anticipatory nausea provides an opportunity to gain greater insight into and greater understanding of the neuroanatomical and neurochemical basis of nausea, potentially leading to new treatments that will boost the endocannabinoid system and thereby reduce or even eliminate nausea.

Since the discovery of the endocannabinoid system, our understanding of the mechanisms by which cannabinoids reduce nausea and vomiting has improved greatly. Animal models demonstrate that CB_1 agonists reduce both acute and anticipatory nausea. CBD and CBDA also show great promise as non-psychoactive treatments that may act by reducing serotonin release in the nausea-inducing regions of the brain. In addition, treatments that boost the endogenous cannabinoid system by inhibiting the degrading enzyme FAAH (thereby elevating AEA), inhibiting the degrading enzyme MAGL (thereby elevating 2-AG), or inhibiting both of those enzymes have shown great promise for treatment of nausea. Since chemotherapy-induced acute and anticipatory nausea are not well controlled with conventional anti-emetic treatments, manipulations of the endocannabinoid system show great promise for reducing that distressing side effect of cancer treatment.

9

Cannabinoids and Pain

The most common medical use of cannabis today and throughout human history is for the treatment of chronic pain, which is usually accompanied by some other disability or mood ·disorder. Estimates indicate that approximately 20 percent of the world's population suffers from chronic pain (Vos et al. 2012). Chronic pain is defined as pain that persists for more than three months. Current pharmacotherapies for chronic pain are ineffective in many patients; approximately 40 percent of patients with chronic pain are not satisfied with their treatment (Corcoran, Roche, and Finn 2015). In the past twenty years, the effectiveness of cannabis in reducing chronic pain has provided increasing evidence that the endocannabinoid system regulates the processing of pain.

Chronic pain can be broadly categorized as consisting of nociceptive pain and neuropathic pain (Beal and Wallace 2016). Nociceptive pain is produced by stimulation of specialized free nerve endings (nociceptors, or "pain receptors") that respond to tissue damage caused by intense chemical (e.g., chili peppers), mechanical (e.g., pinching), or thermal stimulation. Nociceptive pain is pain resulting from actual tissue damage or potential tissue damage; examples include post-operative pain, osteoarthritis-related pain, and mechanical low back pain. Once stimulated, the nociceptor sends a signal to the brain via the spinal cord, triggering a variety of autonomic responses that result in the subjective experience of pain. Neuropathic pain is pain that results from damage to the nervous system itself—examples include diabetic peripheral neuropathy, post-stroke pain and post-therapeutic neuralgia such as chemotherapy-induced neuropathy (Beal and Wallace 2016). Historically, opiates have been the primary class of medications used to treat patients with both acute and chronic pain. Acute pain is well controlled by available therapies, but chronic pain is often resistant to conventional pharmacotherapies. In addition, multiple studies have revealed the adverse consequences of chronic opiate therapy in people without cancer (Rosenblum, Marsch, Joseph, and Portenoy 2008). Therefore, there is an interest in alternative medications for the management of chronic pain. Cannabinoids may have a role in such treatment of chronic pathological pain, including neuropathic pain.

Pre-Clinical Evidence of Analgesic Effects

There is very good pre-clinical evidence that cannabinoids are highly effective in reducing pain in animal models. The tests used to determine the analgesic properties of cannabinoids include tests that model both acute and chronic pain. The simplest tests measure the effect of drugs on acute pain, with analgesic drugs increasing the threshold to detect painful stimuli. In the tail-flick test, a beam of light is focused on a small segment of a rodent's tail; the initial low intensity of the light beam is increased incrementally until the normal pain threshold is reached and the tail is reflexively flicked out of the beam. The hot plate test is equally reliable as an assessment for analgesic drugs using thermal heat. A metal plate at the base of a cylinder is maintained at a temperature between 52° C and 58° C. When a rodent is placed on the hot plate, the latency to respond (by licking a paw or jumping) is measured.

The especially imperative need for new treatments for chronic pain requires the use of pre-clinical models of inflammatory and neuropathic pain to evaluate the potential effectiveness of cannabinoids in relieving distress. Models of chronic pain include models of inflammatory pain and models of neuropathic pain. One model of inflammatory pain is the plantar test, in which an irritant (carrageen, capsaicin, formalin, or lypopolysaccharide) is injected into a rodent's paw and results in a slowly developing inflammation. The paw becomes hypersensitive, and mechanical, thermal, and cold stimuli that normally would not provoke pain become painful (called allodynia); the latency to withdraw the paw from the stimulus is then measured. Analgesic drugs increase the latency for paw movement, and anti-inflammatory drugs reduce edema in the paw. Neuropathic pain models include chemotherapy-induced neuropathy (which produces allodynia) and constriction of the sciatic nerve; analgesic drugs reduce allodynia, as is evidenced by increased latency to respond. Pre-clinical evidence suggests that use of phytocannabinoids may be one of the most promising therapies for chronic neuropathic pain. Not only is THC effective in treatment of chronic pain; pre-clinical evidence suggests that CBD, and perhaps other constituents of cannabis, may also be effective.

THC

Pre-clinical animal models have demonstrated that THC is effective in reducing pain in animal models (Costa and Comelli 2014). Indeed, hyper-algesia is one of the classical "mouse tetrad symptoms" used to classify a drug as acting like THC (Martin et al. 1991). THC has been tested in a wide range of anti-nociceptive assays and has been found to be effective in both acute (e.g., hotplate, tail flick) and chronic (e.g., inflammation, neuropathic pain) models. THC is effective when administered orally, systemically, or directly into the brain or the spinal cord (Costa and Comelli 2014). However, the psychoactive side effects of THC limit its usefulness in treating pain. One approach to overcoming that obstacle is the addition of CBD to THC. CBD has anti-nociceptive effects. Since THC is a mixed CB_1/CB_2-receptor agonist, another approach is the development of CB_2 agonists that only act on the non-psychoactive CB_2 receptors that are primarily located in the peripheral nervous system. Evaluation of THC in pathological pain (such as chronic inflammatory pain, in which the CB_2 receptor plays a pivotal role) has shown it to be an effective anti-nociceptive agent. For instance, in a rat model of chronic arthritic pain, THC was equally potent and effective in non-arthritic rats and arthritic ones. However, in arthritic rats the anti-nociceptive effects of THC were produced via activation of both the CB_1 receptor and the CB_2 receptor, whereas in non-arthritic rats the anti-nociceptive effect was mediated only by its action on the CB_1 receptors (Cox et al. 2007). These data suggest that chronic pain experienced by arthritics involves both peripheral CB_2 receptors and central CB_1 receptors. Therefore, treatments aimed at stimulating peripheral CB_2 receptors may reduce chronic arthritic pain. Another way to reduce the psychoactive side effects of central stimulation of CB_1 receptors is to develop peripherally restricted cannabinoid-receptor agonists that do not cross the blood-brain barrier. Ajulemic acid, a peripherally restrictive analogue of a metabolite of THC, binds to CB_1 receptors and to CB_2 receptors and reduces pain in chronic neuropathic and inflammatory animal models mediated by action at CB_1 receptors only (Dyson et al. 2005; Costa and Comelli 2014).

Because of interactions between the endocannabinoid system and the opiate system, synergistic effects have been reported between THC and opiates in the regulation of pain. Low doses of THC have been found to significantly enhance morphine-induced analgesia when THC and opiates are co-administered systemically into the spinal cord or directly into the ventricles of the brain in animal models. In results from studies in which they are co-administered systemically by injection or orally, a clear synergy has been reported: THC enhanced the anti-nociceptive effects of both morphine and codeine. Therefore, pre-clinical findings suggest that combined treatment with cannabinoids and opioids may be able to produce long-term anti-nociceptive effects at doses that do not produce side effects, without tolerance to each effect (Costa and Comelli 2014).

CBD

Initial investigations of the analgesic effects of CBD in several animal models of acute pain (e.g., hot plate) revealed no evidence for an effect of CBD on pain (Sofia, Vassar, and Knobloch 1975). On the other hand, CBD was subsequently found to be effective in a model of chronic pain when given orally (Formukong, Evans, and Evans 1988). These conflicting results suggested that CBD was not effective against acute pain but was effective against chronic pain. Subsequently, the analgesic properties of CBD were tested specifically in models of persistent and inflammatory pain. Costa et al. (2004) have since found that giving very low oral doses of CBD (15 mg/kg) one hour before carrageenan reduced paw edema within three hours. Even lower doses of CBD (5–7.5 mg/kg, oral) administered two hours after carrageenan-induced inflammation reduced the pain behavior, and higher doses (10–40 mg/kg) eliminated pain behavior and edema. These effects were found to be mediated by the action of TRPV1 receptors as the TRPV1-receptor antagonist (capsazepine) reversed CBD's attenuation of pain (Costa, Giagnoni, Franke, Trovato, and Colleoni 2004). In addition, CBD has been shown to be effective in relieving neuropathic chemotherapy-induced pain in rats, and in diabetic mice (Ward et al. 2014), without the development of tolerance. An additional benefit of CBD is that it has been shown to block the progression of arthritis in a mouse model of collagen-type-II-induced arthritis (Malfait et al. 2000); the synovial (joint) cells from mice treated with an optimal dose of CBD (5 mg/kg, ip, for ten days) released significantly less TNFα, which suggested that the therapeutic effect of CBD on arthritis may be suppression of TNFα, a pro-inflammatory cytokine known to be a major mediator of arthritis.

The molecular targets of CBD that could be responsible for the analgesic effects are numerous, including not only TRPV1 agonism but also agonism of 5-HT$_{1A}$, antagonism of GPR55 receptors, weak antagonism at

CB$_1$ and CB$_2$ receptors, and positive allosteric modulation of glycine receptors. (See Pertwee 2009.) There is evidence for a role of each of these mechanisms in the analgesic effects of CBD. For instance, a recent report suggests that CBD suppresses neuropathic pain in rats with ligation of the spinal nerve by targeting the α3 glycine receptor (Xiong et al. 2012)—mice lacking the α3 glycine receptor, but not CB$_1$ or CB$_2$ receptors, were resistant to the analgesic effect of CBD. More recently, the CBD-induced antagonism of the GPR55 receptor has been shown to play a role in nociceptive signaling. That receptor is highly expressed by mouse primary sensory neurons, and GPR55 knockout mice do not develop hyperalgesia in response to inflammatory or neuropathic stimuli. CBD may also desensitize TRPA1, TRPV1, and TRPV2 channels, resulting in a reduction in hypersensitivity to thermal, chemical, and mechanical stimuli associated with neuropathies (Costa and Comelli 2014). Clearly, there is pre-clinical evidence that CBD attenuates chronic pain through several mechanisms. Because CBD is not psychotropic, it is a strong candidate for treatment of chronic inflammatory and neuropathic pain.

Other cannabis constituents

In addition to THC and CBD, other cannabinoids extracted from cannabis have been shown to have promise as potential analgesic treatments. For example, THCV behaves as a CB$_2$ partial receptor agonist *in vitro* and as an antagonist of the CB$_1$ receptor *in vitro* and *in vivo* (Pertwee 2008). The profile is of interest to pain researchers because there is evidence that CB$_1$ antagonists can reduce pain hypersensitivity by blocking the constitutive activity of the CB$_1$ receptor that maintains sensitized TRVP1 receptors and because CB$_2$ agonists are analgesics without psychoactive effects. Cannabigerol, which has not been well studied, may reduce pain by acting as a α$_2$-adrenergic agonist. And there is evidence that cannabichromene may potentiate the analgesic effect of THC, perhaps by inhibiting re-uptake of AEA. (See Costa and Comelli 2014 for a review.) There have been no clinical studies of the use of CBD alone to treat pain in humans.

GW Pharmaceuticals pioneered the development of nabiximols (whole extracts of cannabis) for therapeutic purposes, most notably for pain in patients with multiple sclerosis. They developed Sativex, which contains THC/CBD ins a ratio of approximately 1 to 1. Recent clinical trials revealed that Sativex is effective (with few side effects at low to moderate doses) in treating pain associated with multiple sclerosis, advanced cancer pain in opiate-resistant patients (Portenoy et al. 2012), and peripheral neuropathic pain accompanied with alloynia (Serpell et al. 2014). Indeed, in Canada Sativex is approved for prescription for pain in MS and cancer and for spasticity in MS.

Such whole-plant extracts contain a mixture of natural cannabinoids and other non-cannabinoid compounds that may interact synergistically to reduce pain. Recent evidence that a high dose of CBD potentiated the ability of a sub-threshold dose of THC to reduce acute pain in the mouse tail-flick test (Varvel et al. 2006) suggests a synergistic effect of CBD and THC in pain reduction. More recently, Comelli et al. (2008) reported that a CBD and THC extract produced a greater analgesic effect than equivalent respective doses of each compound separately in an animal model of neuropathic pain. To determine whether the improved effect was due entirely to THC and CBD or whether it may have been modulated by the other compounds in the extracts, the effectiveness of a combination of the pure compounds was compared against that of the extracts. The pure compounds produced a weaker effect than the plant extracts at the same doses of THC and CBD (Comelli, Giagnoni, Bettoni, Colleoni, and Costa 2008). Therefore, it is necessary to characterize these constituents more fully. Indeed, there is recent evidence that terpenoids present in the plant may facilitate the analgesic effects of the extracts. The terpenoid β-caryophyllene, the most common terpenoid in cannabis, displays anti-inflammatory and analgesic effects by acting as a CB_2 agonist (Gertsch et al. 2008).

Manipulations of the Endocannabinoid System: Pre-Clinical Evidence

Systemic administration of THC and synthetic CB_1-receptor agonists is well known to produce analgesia in animal models of acute and chronic pain. However, concerns about dependence, tolerance, and the cognitive side effects produced by global agonism of the CB_1 receptor and medicinal marijuana remain. Therefore, pre-clinical research has focused on potential manipulations of the endocannabinoid system that do not produce the side effects produced by CB_1 activation. These potential treatments include CB_1 allosteric modulators, FAAH inhibitors, MAGL inhibitors, and CB_2 agonists.

CB_1 allosteric modulators

Endocannabinoids inhibit pain transmission by acting on CB_1 receptors at central, spinal, and peripheral synapses. The use of THC and other CB_1 agonists as analgesics is limited by the psychoactive side effects produced by their action on the orothosteric pocket on the CB_1 receptors throughout the brain. Although endocannabinoids also bind to the orthosteric sites of the CB_1 receptor, both AEA and 2-AG are released on demand where and when they are needed and are quickly metabolized. Therefore, their action is more transient and selective with highly specific temporal and spatial regulation. Allosteric modulators of the CB_1 receptor bind to a distinct site apart from the orthosteric site and produce conformational changes in the receptor, thereby altering the potency of the ligand when it binds to the receptor (Kenakin 2013). However, allosteric modulators have no physiological effect in the absence of ligand binding. Therefore, CB_1-positive allosteric modulators would be expected to enhance the pain-relieving effects of endocannabinoids, but with limited side effects. Indeed, a recent study found that the CB_1 positive allosteric modulator ZCZO11 reduced neuropathic pain and inflammatory pain in pre-clinical animal models without development of tolerance or occurrence of psychoactive side effects (Ignatowska-Jankowska et al. 2015).

FAAH and MAGL inhibitors
Exogenous administration of AEA or 2-AG systemically is unsuitable as a treatment for pain because they are rapidly degraded. However, treatment with inhibitors of the enzymes that degrade these endocannabinoids— MAGL (which degrades 2-AG) and FAAH (which degrades AEA and other fatty acids)—is promising. Because of their "on-demand" production and release, endocannabinoids are specifically generated at sites of nociceptive activity, which prevents unwanted effects of global CB_1-receptor agonism. This strategy also has the potential to improve our current understanding of the functional roles of endogenous AEA and 2-AG in regulating pathological pain. The beneficial effects of activating the endocannabinoid system in different neuropathic pain models were reviewed by Guindon and Hohmann (2009).

Animals genetically engineered to lack FAAH have abnormally high levels of AEA. Although FAAH knockout mice act relatively normal, they are less responsive to pain (Lichtman, Shelton, Advani, and Cravatt 2004). Systemic administration of MAGL inhibitors (Ignatowska-Jankowska et al. 2014; Kinsey et al. 2009; Kinsey et al. 2013; Long, Li, et al. 2009) or of FAAH inhibitors (Fegley et al. 2005; Jayamanne et al. 2006; Kathuria et al. 2003; Lichtman et al. 2004) has been shown to be anti-nociceptive in models of acute and chronic pain, including inflammatory pain and neuropathic pain. Both FAAH inhibition and MAGL inhibition prevent chemotherapy-induced mechanical and cold allodynia (Guindon, Lai, Takacs, Bradshaw, and Hohmann 2013). Both a CB_1 antagonist and a CB_2 antagonist blocked the effects of elevated AEA and 2-AG, but a TRPV1 antagonist blocked only the effect of elevated AEA. In addition, local peripheral (intra-plantar) injection of MAGL inhibitors into the paw of a rat increased local 2-AG levels and blocked pain behavior produced by intra-plantar injections of capsaicin or formalin by a CB_1 and CB_2 mechanism of action of 2-AG (Desroches, Guindon, Lambert, and Beaulieu 2008; Guindon, Guijarro, Piomelli, and Hohmann 2011).

Although FAAH and MAGL inhibitors consistently attenuate neuro-pathic and inflammatory pain in pre-clinical models, they often lack full efficacy in such models, and thus their clinical development is limited. On the other hand, the dual FAAH-MAGL inhibitor JZL195 produced enhanced anti-nociceptive effects in several pre-clinical pain models (Anderson, Gould, Torres, Mitchell, Vaughn, et al. 2014; Long, Li, et al. 2009); however, JZL195 also produced hypomotility, catalepsy, and THC-like subjective effects in a drug-discrimination assay, and impaired spatial memory in the Morris water-maze task in mice (Anderson et al. 2014; Long, Li, et al. 2009; Wise et al. 2012). Although these THC-like psychoactive effects present limitations in the use of dual inhibitors for the treatment of pain, JZL195 was about three times as potent in reducing pain behaviors as it was in producing the psychoactive effects (Anderson et al. 2014). In addition, tolerance is a concern in the development of new analgesics. Tolerance to the analgesic effects of sustained complete FAAH inhibition did not occur (Schlosburg et al. 2010), but tolerance (and physical dependence) did develop to the analgesic effects of sustained complete MAGL inhibition (Ignatowska-Jankowska et al. 2014; Schlosburg et al. 2010). However, full inhibition of FAAH (with a high dose of the FAAH inhibitor PF3845) and partial inhibition of MAGL (with a low dose of the MAGL inhibitor JZL184—4 mg/kg) were recently reported to produce sustained reduction of pain in models of inflammatory and neuropathic pain, with minimal cannabimimetic side effects and no tolerance (Ghosh et al. 2015). Therefore, full FAAH inhibition combined with partial MAGL inhibition may be a preferred treatment for neuropathic and/or inflammatory pain.

URB937, a new peripherally restrictive FAAH inhibitor that does not cross the blood-brain-barrier and therefore cannot produce unwanted psychoactive side effects, has been shown to block neuropathic and inflammatory pain via action of AEA on peripheral CB_1 receptors;

however, unlike actions of 2-AG, peripheral effects of AEA may not involve CB_2 receptors (Clapper et al. 2010). Both brain-penetrant FAAH inhibitors (Naidu, Booker, Cravatt, and Lichtman 2009) and non-brain-penetrant FAAH inhibitors (Sasso et al. 2012) act in a synergistic manner with COX inhibitors (non-steroidal anti-inflammatory drugs) to attenuate inflammatory pain in mice. Therefore, treatments that boost the activity of either AEA or 2-AG when and where they are needed have potential as therapeutic agents for pain relief.

CB_2 agonists

There is evidence that the analgesic effects of THC and synthetic cannabinoids (including the full agonist CP55,940) are mediated not only by their action at the CB_1 receptor but also by their action at the CB_2 receptor (Deng, Cornett, Mackie, and Hohmann 2015). Since CB_2 agonism does not produce the psychoactive side effects of CB_1 agonism, considerable recent research is focused on the potential of selective CB_2 agonists to reduce pain.

CB_2-receptor expression may be confined to immune cells such as macrophages, lymphocytes, and mast cells in the periphery and astrocytes and microglia in the CNS. However, recent studies have demonstrated CB_2-receptor activity on neurons, but whether such activity is present in the absence of inflammation is controversial. Activation of CB_2 receptors mediates the anti-inflammatory effects of endocannabinoids as well as having a role in the anti-hyperalgesia in inflammatory pain states (Pacher and Mechoulam 2011). The CB_2 receptor plays an important role in pain signaling and may be of particular importance in the development of chronic pain states. It is not involved in acute pain, such as that measured in hotplate tests and tail-flick tests; instead CB_2 mechanisms are detected in animal models of persistent or chronic pain (Guindon and Hohmann 2009). Since activation of CB_2 receptors does not produce the psychoactive side effects of activation of CB_1 receptors, recent findings that CB_2-receptor agonists are effective analgesic treatments have great promise for the development of novel treatments for chronic pain. Indeed, recent evidence indicates that chronic CB_2 activation reversed neuropathic pain without the development of tolerance (Deng et al. 2015). Repeated systemic administration of the selective CB_2 agonist AM1710 suppressed peripheral neuropathy produced by administration of the chemotherapeutic agent paclitaxel in a mouse model by reducing the pro-inflammatory cytokine tumor necrosis factor α. Since tolerance develops after prolonged chronic dosing with CB_1 agonists

(Bass and Martin 2000), it is noteworthy that tolerance to the analgesic effects of the CB_2 agonist does not develop. There is considerable pre-clinical evidence that targeting CB_2 receptors to bypass unwanted central effects associated with CB_1-receptor activation is promising as a therapy for neuropathic pain in which the development of effectiveness-reducing tolerance might not occur.

Endocannabinoid Regulation of Pain in the Central Nervous System

Since cannabis and manipulations of the endocannabinoid system reduce acute and chronic pain, it is clear that the endocannabinoid system regulates pain signaling in the central nervous system. The dorsal horn of the spinal cord plays a major role in the processing of pain. It receives and encodes sensory input from the periphery and integrates the descending signals from the brain. When administered into the spinal cord intrathecally (that is, into the spinal theca), cannabinoids reduce acute pain, an effect mediated by CB_1 receptors. However, in sustained painful stimulation, spinal 2-AG levels gradually increase. The increase corresponds to an increase in activation of glial cells and an upregulation of CB_1 receptors, which accompany resolution of a pain state (Alkaitis et al. 2010). Therefore, it is believed that spinal 2-AG signaling initiated by excessive nociceptive activity negatively modulates the acute pain signal, inhibiting the release of pro-nociceptive neurotransmitters (Woodhams, Sagar, and Chapman 2015).

In chronic pain, such as neuropathic pain resulting from peripheral nerve damage, the spinal nociceptive circuitry becomes sensitized, which results in hyperalgesia (excessive pain after a nociceptive stimulus) and allodynia (perception of a normally innocuous stimulus as painful). Treatments that boost the endocannabinoid system have great potential for alleviating these chronic pain states, whereas current analgesics are ineffective (Woodhams et al. 2015). In animal models of neuropathic pain, both CB_1 receptors and CB_2 receptors are upregulated, as are levels of AEA and 2-AG in the spinal cord. However, in view of the role of 2-AG signaling at CB_2 receptors in resolving pain (Alkaitis et al. 2010), a novel area of ongoing research involves intrathecal injections of CB_2-receptor agonists to treat chronic pain states (Burston et al. 2013). Spinal CB_2 expression is elevated and CB_2 agonism alters spinal nociceptive activity in a model of osteoarthritic pain, without effect on control animals. The CB_2 receptor is primarily expressed on glial cells of the CNS; therefore these cells are most likely to mediate the effects of CB_2 agonists in models of persistent pain. Activation of CB_2 receptors can reduce the release of pro-inflammatory cytokines from glial cells, which is important in the analgesic mechanism.

In the brain, the endocannabinoid system influences ascending pain signals in the thalamus, influences descending modulatory signals in the brainstem, and influences the affective/emotional aspects of pain sensation through actions in the higher cortico-limbic circuits. Direct infusion of cannabinoid agonists into the brainstem regions of the peri-aqueductal gray (PAG) and the rostral ventromedial medulla produces anti-nociceptive effects that can be blocked by CB_1 antagonists. The involvement of endocannabinoids at these sites is mediated by the release of AEA after electrical stimulation of the PAG or after peripheral inflammatory insult. PAG levels of AEA and 2-AG are also elevated in animal models of neuropathic pain. FAAH inhibition in the PAG is anti-nociceptive in acute pain tests, but at very high levels AEA produces pro-nociceptive effects by action at TRPV1 receptors. The PAG is also the site of action of endocannabinoids mediating stress-induced analgesia in rodents (Hohmann et al. 2005). In the region of the rostral ventromedial medulla of the brainstem, "on" cells facilitate nociceptive activity, whereas "off" cells inhibit nociceptive activity; the action of endocannabinoids in that region inhibits the "on" cells and promotes the firing of the "off" cells (Woodhams et al. 2015).

Higher brain regions in the limbic system (in particular the amygdala) and cortical regions mediate the emotional components of pain. Indeed, pre-clinical studies demonstrate that lesions of the amygdala (which regulates emotional behavior) reduce the analgesia produced by THC. A recent fMRI study of healthy humans (Lee et al. 2013) investigated the effects of THC on brain activity produced by ongoing cutaneous burning pain induced by topical administration of capsaicin (an alkaloid derived from chili peppers). Lee et al. (ibid.) found that THC reduced the reported unpleasantness, but not the intensity, of pain. The reduced unpleasantness of pain was accompanied by reduced activity in the anterior cingulate cortex (a cortical region involved in the perception of pain) and by reduced functional connectivity between the amygdala and primary sensorimotor areas during the pain state. Interestingly, the reduction in connectivity was positively correlated with the reductions in ratings of unpleasantness of pain. These findings suggest that the amygdala activity may be related to inter-individual differences in response to cannabinoid analgesia.

Clinical Trials in Humans

There are few effective therapeutic options for patients living with chronic pain. It has been reported that only 40–60 percent of patients obtain even partial relief of their pain with current medications (Wilsey et al. 2008). There is very good evidence that cannabinoids, including marijuana, can be effective in relieving pain as assessed on a visual analog scale, even at low doses (1.29 percent THC) that do not produce psychoactive or cognitive side effects (Wilsey et al. 2013). There have been no clinical trials with humans of the efficacy of CBD alone for chronic pain.

Mary Lynch and colleagues (Lynch and Campbell 2011; Lynch and Ware 2015) conducted systematic reviews of randomized controlled trials conducted since 2003 examining cannabinoids in the treatment of chronic non-cancer pain. The cannabinoids included in the reviews were smoked cannabis, oromucosal extracts of cannabis-based medicines, nabiximols, nabilone, dronabinol, and a novel THC analogue of the metabolite THC-11-oic acid (ajulemic acid). The chronic pain conditions included neuropathic pain, fibromyalgia, rheumatoid arthritis, and mixed chronic pain. Among the 29 trials included in the reviews, 22 found significant analgesic effects of the cannabinoid as compared with either placebos or active control compounds, and several reported significant improvements in sleep. No serious adverse effects were reported; the adverse effects that were reported were mild to moderate and led to only a few participants withdrawing from the studies. The reviews concluded that there is evidence that cannabinoids are safe and moderately effective in treating neuropathic pain, and that there is some evidence of efficacy in fibromyalgia and rheumatoid arthritis. Of particular importance, two of the trials examining smoked cannabis (Abrams et al. 2007; Ellis et al. 2009) demonstrated a significant analgesic effect in HIV neuropathy, a type of pain highly resistant to any available treatment for neuropathic pain. In addition, Abrams et al. (2011) have shown that vaporized cannabis can augment the analgesic effects of opioids, an effect also demonstrated in the pre-clinical animal literature; it will be interesting to determine if this important finding can be replicated using a double-blind placebo control procedure. It was noteworthy that in a trial of the effect of Sativex in the treatment of rheumatoid arthritis a significant reduction in disease activity was also noted (Blake et al. 2006), which is consistent with pre-clinical work showing that cannabinoids are anti-inflammatory.

High doses of THC may not be necessary to control neuropathic pain. A recent double-blind, placebo-controlled crossover study (Wilsey et al. 2013) evaluated the potential of vaporized cannabis (placebo, medium-dose THC [3.53 percent], or low-dose THC [1.29 percent]) to reduce neuropathic pain in humans who were resistant to traditional treatment. Both the high and the low dose were equally effective in reducing pain, yet the psychoactive effects of the low dose were minimal and were well tolerated.

The first cohort study of the long-term safety of medical cannabis use was recently reported by Mark Ware and colleagues; it involved patients being treated for non-cancer chronic pain in seven clinics in Canada over the course of a year (Ware, Wang, Shapiro, Collet, et al. 2015). The primary outcome measures were severe adverse events and non-severe adverse events as defined by the International Conference on Harmonization (http://www.ich.org/cache/compo/276-254-1.html). The comparison groups included 215 individuals with chronic pain who were given cannabis (12.5 percent THC) and 216 individuals with chronic pain who were not given cannabis. The cannabis users were advised to use the delivery method with which they were most comfortable and to titrate their dose to the level they tolerated best. The median intake was 2.5 grams per day, and no relationship was found between increasing the daily dose and the development of adverse effects. This suggests that the patients titrated their dose to control adverse effects. Over the year of treatment the cannabis group, but not the control group, showed a significant reduction in pain intensity. In addition, there was no difference among the groups in the risk of severe adverse events (cannabis group 13 percent, non-cannabis group 19 percent). Most patients in the cannabis group (88.4 percent) and most in the control group (85.2 percent) reported at least one non-severe adverse event, with a mean of three events per participant in the cannabis group and a mean of two events per participant in the control group. The overall incidence of non-severe adverse events in the cannabis group (818) was significantly higher than that in the control group (581). The specific events associated with cannabis included headache, dizziness, nausea, and somnolence.

As a secondary measure, Ware et al. (2015) administered neurocognitive tests at the start of the study, at 6 months, and at 12 months, using two sub-tests of the third edition of the Wechsler Memory Scale (Verbal Paired Associates I, including recall, and Verbal Paired Associates II, including recall and recognition) and two sub-tests of the third edition of the Wechsler Adult Intelligence Scale (Digit Symbol-Coding and Picture

Arrangement). Significant improvements were observed in all neurocognitive sub-tests after 6 months and after 12 months in both the cannabis group and the controls. No differences in neurocognitive function were seen between the two groups. Measures of pulmonary function did not differ between the groups after adjusting for tobacco smoking. The cannabis users did not show changes in liver, kidney, or endocrine functioning as assessed by blood tests. Therefore, this first follow-up study of medical cannabis users suggests that its adverse effects are small and are comparable both quantitatively and qualitatively to those of prescription cannabinoids, such as naboline. The average dose of 2.5 g/day of 12.5 percent THC cannabis may be safe as a part of a carefully monitored pain-management program (Ware et al. 2015).

Using a conservative criterion of risk of bias of results, Whiting et al. (2015) conducted an extensive systematic review of 79 randomized controlled trials that examined the benefits and adverse effects associated with medical cannabis across a broad range of conditions, the majority of the trials evaluating nausea and vomiting due to chemotherapy or chronic pain and spasticity due to MS and paraplegia. In determining the quality of a trial, Whiting et al. used the Cochrane Collaboration's tool for assessing risk of bias in randomized trials, which covers six domains of bias: selection bias, performance bias, detection bias, attrition bias, reporting bias, and other bias. If at least one domain was rated high, a trial was considered at high risk of bias. The review concluded that most studies suggested that cannabinoids were associated with improvements in symptoms. The evidence was strongest that smoked cannabis and oromucosal THC:CBD mixtures may be beneficial for the treatment of chronic neuropathic or cancer pain and that nabiximols, naboline, THC/CBD capsules, and dronabinol may be beneficial for the treatment of spasticity due to multiple sclerosis, with moderate risk of bias for most of these studies. Although the cannabinoids also were effective for chemotherapy-induced nausea and vomiting, the risk of bias among these studies was high. Interestingly there was no clear evidence for a difference in effectiveness or in adverse effects based on the type of cannabinoid or the mode of administration.

Although the use of FAAH and MAGL inhibitors in pre-clinical studies has clearly supported the use of endocannabinoid-targeted compounds in clinical pain trials with humans, there has been only one published report of such a trial, and it was a failure. The pharmaceutical corporation Pfizer developed a highly selective FAAH inhibitor, PF-04457845, to produce analgesia in an osteoarthritic patient population (Huggins,

Smart, Langman, Taylor, and Young 2012). That compound reduced FAAH activity by 96 percent and increased AEA levels substantially, but was not differentiated from placebo in reduction of pain. However, it was well tolerated, with no evidence of cannabinoid-type adverse events.

The pain targeted in the Pfizer clinical trial, osteoarthritis pain, may differ qualitatively from the pain typically measured in many pre-clinical models that have demonstrated analgesic effects of FAAH inhibition. Most pre-clinical models measure reflex responses to a mechanical or thermal stimulus, whereas the predominant symptom in neuropathic pain evident in osteoarthritis is not evoked pain but instead spontaneous pain, which is more difficult to model pre-clinically (Fowler 2015). In an attempt to model osteoarthritic pain in animals more closely, Bryden et al. (2106) injected monosodium iodoacetate into rats' knees, producing histological changes representative of those seen in human osteoarthritic patients. The measure of pain was the likelihood that a rat would spontaneously burrow into bedding (an innate rodent behavior indicative of well-being). Bryden et al. (2016) demonstrated deficits in burrowing in this model that were reversed by COX inhibitors (ibuprofen and celecoxib), but were not reversed by the Pfizer FAAH inhibitor, PF-04457845, which was also ineffective in human osteoarthritis patients (Bryden et al. 2015). The pre-clinical data thus mirror the human clinical data for this indication. In view of the considerable pre-clinical evidence of the potential of FAAH inhibitors and MAGL inhibitors to reduce pain in a variety of models, there is a clear need to continue to evaluate the potential benefits of these treatments in other human models of chronic pain.

Conclusion

A majority of the patients who use medical marijuana are prescribed marijuana for pain—particularly chronic neuropathic pain, which is resistant to current treatments. Since opiates are the most common treatment for pain, it is interesting that considerable pre-clinical animal research demonstrates a clear interaction between cannabinoids and opiates at a number of levels within the cell, including direct receptor associations, alterations in endogenous opiate release, and post-receptor interactions via shared signal transduction pathways. Various studies have demonstrated cross-tolerance, mutual potentiation, and receptor cross-talk between the μ-opiate receptor and the CB_1 receptor. Drugs that target the cannabinoid system often affect the opioid system in tandem.

Considerable evidence indicates a synergistic effect of cannabinoid-opiate drugs (Scavone et al. 2013). For instance, sub-threshold combined doses of cannabinoids and morphine reduce pain in animals and in humans. These findings suggest that the combined use of THC and opiates may provide an opportunity for better pain relief from lower doses of each drug. In addition, the use of nabiximols (Sativex) for chronic pain associated with multiple sclerosis and cancer has shown promise, with the additional benefit that CBD tempers the intoxicating effects of THC.

Although the first clinical trial of a FAAH inhibitor to produce analgesia in an osteoarthritic population failed, the pre-clinical evidence suggests that continued investigation of both FAAH and MAGL inhibitors is warranted, especially for non-osteoarthritic chronic pain. Pre-clinical evidence points to future development of effective CB_2 agonists, devoid of the psychoactive side effects of CB_1 agonists, to produce pain-relieving anti-inflammatory effects.

10

Cannabinoids and Epilepsy

Epilepsy is a neurological disorder that manifests as recurrent, spontaneous seizures or convulsions with possible loss of consciousness due to disturbance of excitatory-inhibitory equilibrium of neuronal activity in the brain. It affects approximately 1 percent of the world's population (Williams, Jones, and Whalley 2014). Two broad categories of seizures have been described: generalized and focal. Generalized seizures originate at a specific point within the brain but rapidly distribute across the brain to affect both hemispheres. Focal seizures are restricted to a specific region of the brain or a single hemisphere.

Conventional anti-epileptic drugs block sodium channels or calcium channels, or enhance GABA function to reduce the release of excitatory glutamate, thereby preventing the spread of the seizure within the brain. Current anti-epileptic drugs are effective in approximately 50 percent of patients. However, 30 percent of the epileptic population experience intractable seizures regardless of the anti-epileptic drug used, and 50 percent of the population will eventually become resistant to currently available treatments. All existing anti-epileptic drugs are associated with numerous side effects (impairment of motor function, cognitive dysfunction, emotional lability). Therefore, there is a need for the development of better treatment options (Williams et al. 2014). Recent evidence shows promise for cannabinoids in the treatment of epilepsy.

Anti-Convulsant Effects of Cannabis

Before the discovery and development of modern anti-epilepsy drugs, cannabis was a drug of choice for treating seizures. In 1840, William O'Shaughnessy described the successful treatment of seizures in an infant by using cannabis tincture (O'Shaughnessy 1840). Queen Victoria's personal physician, J. R. Reynolds, described cannabis as the most useful agent that he was acquainted with to treat violent convulsions (Reynolds 1868). We now know that whole cannabis contains several cannabinoids with diverse pharmacology, so it is difficult to interpret the mechanisms of action of whole cannabis plants (Williams et al. 2014).

There have been no clinical trials of the effectiveness of whole cannabis or THC effects on seizures in epilepsy patients. However, there have been a few case studies of epilepsy patients who are also cannabis users. Caution must be exerted in interpreting these findings, because of positive bias (Williams et al. 2014). According to one report, 24-year-old epileptic required two to five cannabis cigarettes (self-prescribed) in addition to prescribed doses of phenobarbital and phenytoin to remain seizure free (Consroe, Wood, and Buchsbaum 1975). In contrast, another report suggested a pro-convulsant effect of cannabis in which an epileptic patient receiving conventional anti-epilepsy medication was seizure free until engaging in a period of cannabis use (Keeler and Riefler 1967). Grinspoon and Bakalar (1997) reported that two patients who replaced traditional anti-epilepsy drugs with cannabis smoking showed improvement in seizure frequency and severity. An interview-based survey of 295 epilepsy patients treated in a tertiary care center revealed that 21 percent used cannabis and the majority reported beneficial effects on seizures (Gross, Hamm, Ashworth, and Quigley 2004). One pre-clinical animal study investigated the anti-convulsant effects of whole-plant cannabis injected in rats (Ghosh and Bhattacharya 1978), another the effects of whole-plant cannabis inhaled by dogs (Labrecque, Halle, Berthiaume, Morin, and Morin 1978); both of those studies supported an anti-convulsant effect of cannabis. Overall, the evidence supports an anti-convulsant effect of cannabis; however, there have been no well-controlled human studies, and a report by Williams et al. (2014) of potential pro-convulsant effects warrants caution.

Anti-Convulsant Effects of CB₁ Agonists

Most of the evidence that THC and other CB_1 agonists may reduce seizures is found in the pre-clinical animal literature (Chesher, Jackson, and Starmer 1974). When administered orally, doses of 160–200 mg/kg of THC protected against seizures in a mouse epilepsy model (Boggan, Steele, and Freedman 1973), which led Boggan et al. to suggest that the median effective dose for 50 percent of the population (ED50) of oral THC for seizure protection is about 200 mg/kg. At the lower dose of 50 mg/kg, THC was pro-convulsant, but that effect could be reversed by co-administration of CBD (50 mg/kg). Later work showed that administration of THC and the anti-epileptic drug phenobarbitone produced a synergistic anti-convulsant effect that was further enhanced by CBD (Chesher, Jackson, and Malor 1975). THC and a number of synthetic CB_1 agonists have consistently been shown to produce CB_1-dependent anticonvulsant activity in experimental models of seizure and epilepsy. (For a review see Williams et al. 2014.) In addition, experimental studies have demonstrated that antagonism of the CB_1 receptor exacerbates seizure activity in animal models (Blair, Deshpande, and DeLorenzo 2015).

An important case report demonstrating the role of the endocannabinoid system in the regulation of synaptic transmission in controlling epileptic seizures was reported in a patient who enrolled in a trial of the CB_1 antagonist rimonabant for treatment of obesity. The 52-year-old patient, who had a history of adolescent epilepsy, had been seizure free for several years. After administration of rimonabant, the patient experienced new onset of nocturnal partial seizures (Braakman, van Oostenbrugge, van Kranen-Mastenbroek, and de Krom 2009), which then subsided until the re-administration of rimonabant. Since a CB_1 antagonist may be pro-convulsant, this report provided evidence that CB_1 agonism may be a potential anti-convulsant treatment for epilepsy. However, CB_1 agonists produce psychoactive side effects, and prolonged CB_1 activation may result in exacerbation of seizure activity—probably as a result of receptor adaptation (Blair et al. 2015).

Epilepsy and Dysregulation of the Endocannabinoid System

The effectiveness of CB_1 agonists in reducing seizures in pre-clinical models of epilepsy suggests that the endocannabinoid system may be a homeostatic regulator of that disorder. Indeed, CB_1-expressing inhibitory and excitatory synapses act to regulate dynamically changing normal and pathological oscillatory neural-network activity. Endocannabinoids control the activity of these neural networks.

Increasing evidence suggests that epilepsy can modify the endocannabinoid signaling system substantially. Temporal-lobe epilepsy, one of the most common forms seen in adults, is associated with changes in the hippocampus in which CB_1 expression is downregulated during the acute phase, shortly after the precipitating event, but then upregulated in the chronic phase of the disorder (Soltesz et al. 2015). In the hippocampus, CB_1 receptors are present on both inhibitory GABA-containing terminals synapses and excitatory glutamate-containing terminals. As a consequence, the activation of CB_1 receptors on glutamatergic terminals should be anti-convulsive by reducing glutamate release, and the activation of CB_1 receptors on GABA terminals should be pro-convulsive by reducing GABA release. The specific loss of CB_1 receptors (receptor downregulation) on the glutamatergic terminal has been shown to lead to an increased excitability of glutamatergic neurons treated with the glutamate agonist kianic acid and to an increase in seizures induced by that acid (Marsicano et al. 2003). Taken together, evidence from both patients and animal models indicates that, as the disease progresses, concurrent upregulation of CB_1 receptors on GABAergic terminals and downregulation of CB_1 receptors on glutamatergic axons may mechanistically contribute to seizures. However, the relative importance of the upregulation and downregulation of CB_1 receptors at GABAergic versus glutamatergic terminals is not well understood (Soltesz et al. 2015). For example, data from cell-type-specific knockout mice indicates that the selective deletion of CB_1 on excitatory glutamate cells, but not on inhibitory GABAergic interneurons, worsens kainate-induced acute seizures (Monory et al. 2006).

Epilepsy is characterized by recurrent, unprovoked seizures. Temporal-lobe epilepsy can be studied experimentally in rodents. Chemical convulsant agents (kianate and pilocarpine) produce intermittent or continuous acute seizures, which can last up to a few hours. These are followed by a latent period that can be interrupted days or weeks later by spontaneous chronic epileptic seizures. Data from such experimental models of epilepsy suggest that excessive glutamate release from axon terminals during seizures may trigger negative-feedback inhibition of further glutamate release through the endocannabinoid signaling system (Katona and Freund 2012). Action of the endocannabinoid system dampens overactive circuits. Indeed, the level of endocannabinoids is increased in the brain shortly after it is subjected to acute insults (Hansen, Schmid, et al. 2001; Wallace, Blair, Falenski, Martin, and DeLorenzo 2003). These transient surges in endocannabinoids in the otherwise normal (that is, not chronically epileptic) brain engage pre-synaptic CB_1 signaling and exert an overall protective effect against over-excitation (Soltesz et al. 2015). However, in individuals with chronic ongoing epilepsy this endocannabinoid-mediated negative-feedback control system seems to be functionally compromised, either because of alterations in synthesis pathways or because of decreased CB_1 expression on glutamatergic terminals, and is therefore unable to prevent the generation of seizures.

Indeed, in the chronic phase of temporal-lobe epilepsy, the levels of AEA and the 2-AG synthesizing enzyme DAGL, without changes in the level of the 2-AG degrading enzyme MAGL, are reportedly decreased in the cerebrospinal fluid or in the brain of individuals with temporal-lobe epilepsy (Ludanyi et al. 2008; Romigi et al. 2010). Thus, diminished production of endocannabinoids may contribute to epileptic seizures. Thereby, CB_1 agonists may serve as anti-convulsants by decreasing glutamatergic transmission. Even in chronically epileptic animals, CB_1 antagonists increase the frequency and the duration of seizures (Soltesz et al. 2015; Wallace et al. 2003).

Promising new treatments based on MAGL inhibition came to light in pre-clinical animal studies in which the kindling model of epilepsy was used. Kindling was induced by electrically stimulating the amygdala at a frequency and an amplitude known to induce seizure activity in mice. The highly selective MAGL inhibitor JZL184 raised seizure thresholds, delayed the development of generalized epileptic seizures, and decreased the number of seizures after discharge duration in the kindling model of temporal-lobe epilepsy. However, once mice were fully kindled, MAGL inhibition no longer had an anti-convulsant effect. This effect was mediated by the action of 2-AG on CB_1 receptors located on glutamatergic neurons, because JZL184 was ineffective in conditional CB_1 knockout mice on glutamatergic forebrain neurons. A less selective MAGL inhibitor, URB602, also was found to increase the latency of general seizures induced by pentylenetetrazole (PTZ) in rats (Naderi, Ahmad-Molaei, Aziz Ahari, and Motamedi 2011). Previous work had demonstrated that the CB_1 agonist WIN 55212,2, but not the FAAH inhibitor URB597, also delayed the development of generalized seizures in the model, but not when mice were fully kindled (Wendt, Soerensen, Wotjak, and Potschka 2011). These findings suggest that 2-AG, not AEA, is the endocannabinoid that modulates seizures, and that it does so by reducing glutamate release during the development of seizures in kindled mice. It is also known that CB_1 receptors on glutamatergic neurons can mediate a protection against excitotoxic seizures (Bhaskaran and Smith 2010; Monory et al. 2006; Ruehle et al. 2013). This indirect activation of CB_1 receptors for treatment of seizures offers a clear advantage over global CB_1-receptor activation. Indeed, *in vitro* studies show that 2-AG reduces excitatory post-synaptic currents in hippocampal slices from mice with temporal-lobe epilepsy (Bhaskaran and Smith 2010). However, caution must be exercised in interpreting the results, because Ma et al. (2014), using a much higher dose of JZL184 (20 mg/kg twice a day rather than 8 mg/kg once a day), found that chronic JZL184 *increased* seizure frequency in a mouse model of temporal-lobe epilepsy. This could be a biphasic effect of cannabinoids with high doses producing an opposite effect than low doses, or it could be due to the development of tolerance mediated by downregulation and desensitization of CB_1 receptors after chronic cannabinoid administration. Further studies in animal models with spontaneous recurrent seizures are needed to determine the generality of these effects.

Anti-Convulsant Effects of CBD

CBD, which does not act on CB_1 receptors, is the only isolated phytocannabinoid to have been investigated for anti-convulsant effects both pre-clinically in animals and clinically in humans. The first pre-clinical evidence for the anti-convulsant effects of CBD was published by Izquierdo, Orsingher, and Berardi (1973). CBD (1.5–12 mg/kg, ip) given one hour before induction of seizures in a mouse model reduced the severity relative to controls. Ralph Karler and colleagues (Karler, Cely, and Turkanis 1973; Karler and Turkanis 1978) subsequently demonstrated the anti-convulsant effects of orally administered CBD (120 mg/kg) and THC (100 mg/kg) in mice. Karler and colleagues also showed that CBD (0.3–3 mg/kg, ip) increased the electrophysiologically recorded epileptic after-discharge threshold in electrically kindled limbic seizures in rats, as also did the anti-epileptic drug phenytoin. However, CBD went beyond phenytoin by also reducing the after-discharge amplitude, duration, and propagation (Turkanis, Smiley, Borys, Olsen, and Karler 1979); the authors suggested that CBD was the most efficacious of the drugs tested against limbic after discharges and convulsions.

CBD clearly shows anti-convulsant effects in acute pre-clinical models of epilepsy, but the mechanism of action is not well understood. It is possible that CBD may reduce neuronal excitability and neural transmission by modulation of intracellular calcium through interaction with TRPV1 receptors or GPR55 receptors (Devinsky et al. 2014). Indeed, the latter mechanism is particularly interesting. GPR55 receptors are localized on excitatory axon terminals, where they facilitate glutamate release when the neuron fires. Because CBD effectively blocks GPR55 activation, it would be an ideal candidate as an anticonvulsant agent. Thus CBD may selectively dampen excess presynaptic glutamate release from only the hyperactive excitatory neurons during epileptic seizures (Sylantyev et al. 2013; Katona 2015).

CBD may not be as effective in animal models of chronically epileptic animals. Colasanti, Lindamood, and Craig (1982) used cortical implantation of cobalt to model chronic seizures in humans and found that CBD (60 mg/kg, ip) was ineffective. On the other hand, recent work has verified that CBD shows significant anti-epileptiform and anti-convulsant activity in a wide variety of *in vitro* and *in vivo* models (Jones et al. 2010). Cannabidivarin, the propyl variant of CBD, has recently been shown to have anti-convulsant properties in the same models (Hill, Mercier, Hill, Glyn Jones, Yamaksi, et al. 2012).

Human clinical trials with CBD are ongoing, and there is interest in its use in childhood epilepsy. The psychotropic effects of THC limit or prohibit widespread therapeutic use—particularly in epilepsy, for which regular, repeated, lifelong dosing is necessary. However, not only do all licensed anti-epileptic drugs exert significant motor or cognitive side effects (Fisher 2012); because of side effects, poorly controlled seizures, or a combination of the two, many epilepsy patients cannot drive or maintain employment. In pre-clinical animal research, CBD has been reported to be a more reliable anti-convulsant than THC, and without THC's psychoactive or motor side effects. Therefore, CBD may be a better therapeutic option for epilepsy than THC.

In the past five years there has been great public interest in the use of "medical marijuana" products for the treatment of pediatric epilepsy. Sanjay Gupta of the television channel CNN has produced two hour-long specials on the efficacy of CBD-rich cannabis in treating Dravet Syndrome, a severe form of childhood epilepsy that is usually resistant to standard anti-epilepsy drugs. Beginning in the second year of life, children affected by Dravet Syndrome develop an epileptic encephalopathy that results in cognitive, behavioral, and motor impairment (Devinsky et al. 2014). Thus, early and effective therapy is necessary. The increase in public interest is attributable to reports that two young children with Dravet Syndrome stopped having seizures after taking CBD-rich marijuana preparations. These remarkable anecdotal stories have produced high expectations regarding the therapeutic potential of CBD for the treatment of epilepsy. Families have petitioned legislators for access to CBD-rich strains of marijuana, and some families have moved to states with more liberal marijuana policies in order to gain access to products for their affected children. Physicians may be caught in a quandary complicated by insufficient scientific data.

Nearly 40 years ago, on the basis of pre-clinical animal research indicating the effectiveness of CBD in reducing epileptic seizures, Raphael

Mechoulam and his colleagues provided the first scientific clinical trial of the potential of CBD to reduce epileptic seizures in a small group of patients. Mechoulam treated patients in Israel daily with 200 mg CBD ($n = 4$) or placebo ($n = 5$) (Mechoulam and Carlini 1978). While the placebo-treated patients showed no improvement during the three-month trial, three of the four patients treated with CBD showed a major reduction in seizure frequency (ibid.). Subsequently, a research group in Brazil (Cunha et al. 1980), in collaboration with Mechoulam, used a double-blind procedure to randomly assign a total of 15 epilepsy patients to receive high doses (200–300 mg/day) of CBD ($n = 8$) or placebo ($n = 7$) in capsules for 4.5 months, along with the anti-epilepsy drugs pre-scribed before the experiment. Among the eight patients treated with CBD, only one showed no improvement. Four were nearly free of con-vulsions throughout the experiment, and another three showed partial improvement. The patients reported no severe adverse effects from CBD. Despite these initial promising results, which were consistent with the pre-clinical animal literature, it is only very recently that subsequent clin-ical trials have been conducted, largely in response to the large number of anecdotal reports of the positive effects of CBD-rich marijuana on children with pediatric epilepsy. As Mechoulam (2015) asks, why did we have to wait decades?

Even though the efficacy of CBD for treating epilepsy in humans was supported nearly 40 years ago, until recently the published data on the use of CBD for the treatment of epilepsy have come from fewer than 70 participants, very few of them children (Ames and Cridland 1986; Trem-bly and Sherman 1990). Few of the studies were rigorous, and few of them used high-quality evidence (Whiting et al. 2015). More recently, Porter and Jacobson (2013) published self-reports of the experiences of parents who had given their children (nineteen in all) some form of high-CBD product for severe intractable epilepsy. The reported doses of CBD ranged from below 0.5 mg/kg/day to about 30 mg/kg/day. The majority of the families reported improvement, ten of them reporting greater than 80 percent improvement and two reporting complete cessation of sei-zures. Others reported having discontinued other medications (Porter and Jacobson 2013). Although this does not constitute a high-quality placebo-controlled experiment, it does provide some information that was missing until 2013.

In view of the current interest in CBD as a treatment for childhood epilepsy, pharmaceutical companies are developing CBD products for treating Dravet Syndrome and other forms of epilepsy and are evaluating their efficacy. GW Pharmaceuticals has developed a CBD-rich compound, called Epidiolex, that, like Sativex, is administered by sublingual spray. Epidiolex has received both an Orphan Drug Designation and a Fast Track Designation from the US Food and Drug Administration for the treatment of Dravet Syndrome. Initial reports of several children treated with Epidiolex for intractable epilepsy were promising (Devinsky et al. 2016); however, the initial study lacked placebo controls. Epidiolex reduced overall seizure frequency by 54 percent in all patients and by 63 percent in Dravet Syndrome patients. After 3 months of treatment, 9 percent of all patients and 16 percent of Dravet patients were seizure free. Adverse effects (somnolence, diarrhea, fatigue and decreased appetite) were not severe. Therefore, the limited data based on self-reported seizure frequencies are promising, although placebo controls were absent.

On March 14, 2016, GW Pharmaceuticals issued a press release announcing positive Phase 3 clinical trial results for Epidiolex (CBD) for the treatment of Dravet Syndrome. In the study, 120 patients were randomly assigned to receive 20 mg/kg/day of Epidiolex ($n = 61$) or placebo ($n = 59$). Epidiolex was added to the current treatments the patients were receiving for epilepsy (which, because of failure to provide relief of seizures, consisted of an average of three different drugs). The mean age of patients was 10 years, and 30 percent of the patients were under 6 years of age. The median baseline frequency of convulsive seizures was 13 per month. The primary endpoint compared Epidiolex and placebo on the percentage change in the monthly frequency of convulsive seizures during the 14-week treatment period, with a 4-week baseline observation period. The results, which were highly statistically significant ($p = 0.01$), showed that Epidiolex achieved a median reduction in monthly number of convulsive seizures of 39 percent, versus 13 percent for placebo. The difference between Epidiolex and placebo was evident during the first month of treatment and was maintained for the entire treatment period. Epidiolex was generally tolerated well, the most common adverse events being somnolence, diarrhea, decreased appetite, fatigue, pyrexia, vomiting, lethargy, infection of the upper respiratory tract, and convulsion. Of the patients who reported an adverse event, 84 percent reported it to have been mild or moderate. Eight patients on Epidiolex discontinued treatment because of adverse effects; only one patient on placebo did so. This is very promising placebo-controlled evidence in support of both the safety and the efficacy of CBD in treatment of a severe and difficult-to-treat form of epilepsy that is devastating to very young children.

A number of families in the United States and in Canada have gained access to various "hemp oil" preparations with high CBD and low THC content. Assessing the outcomes of the use of such preparation is difficult because of high variability between products, lack of consistency in dosing, variable quality control, and uncertainty about the presence of other potentially active cannabinoids. In a recent report from Colorado, although 57 percent of families using such products reported positive results, there was no evidence of improvement in EEG pattern in eight of the patients who reportedly responded to the drug. And there were significant adverse effects, including increased seizures in 13 percent of the patients (Press, Knupp, and Chapman 2015). Interestingly, a higher rate of benefit was reported by families that had moved to Colorado specifically in order to gain access to "hemp oil" products than by families who had already been living there, which suggests a placebo effect in self-reported outcomes.

Obtaining "evidence-based medicine" becomes complicated when the substance in question (e.g., CBD) is controlled by rigid and intimidating federal regulations and by rapidly changing state regulations. According to the US Controlled Substance Act of 1970, marijuana and THC are classed under the federal government's Schedule I of controlled substances—a classification suggesting that they have a high potential for abuse, that they have no currently accepted medical use, and are not accepted as safe. The Controlled Substance Act of 1970 specifies that all species of plants and substances derived from marijuana fall under Schedule I. Therefore, CBD is subject to the same restrictive laws as THC, although "most experts and considerable evidence now suggest that this particular phytochemical in fact does not have abuse potential and is clearly of substantial medical interest" (Filloux 2015). Some states have enacted CBD-specific laws that allow families to administer CBD-rich, THC-poor marijuana products to children with intractable epilepsy, but the families must determine how to obtain the substances and must determine how large a dose to administer. Indeed, under federal law it is technically illegal for a physician to prescribe these substances. Because of the therapeutic potential of CBD for epilepsy, the American Academy of Pediatrics has recently published a position paper stating its position against legalization but favoring changing the status of marijuana from Schedule I to Schedule II to facilitate high-quality scientific research.

11

Cannabinoids and Neurodegenerative Disorders

In a rigorous meta-analysis published in the *Journal of the American Medical Association* under the title "Cannabinoids for medical use," Whiting et al. (2015) concluded that the best evidence in the clinical human literature for medical use of cannabinoids pertained to chronic pain and spasticity in multiple sclerosis (MS). Recently nabiximols (Sativex) has been approved for prescription for MS spasticity in Canada and in several other Western countries. There is evidence that cannabinoid-based medicines may alleviate not only symptoms of spasticity in MS but also symptoms of other neurodegenerative disorders.

Numerous pre-clinical studies have addressed the ability of cannabinoids to protect not only neurons but also some glial cell sub-populations from various insults. This suggests that cannabinoids may delay or arrest disease progression in neurodegenerative disorders. It is known that cannabinoids reduce excitatory glutamate release, reduce intracellular calcium, produce anti-oxidant effects, and produce anti-inflammatory effects—all neuroprotective effects. Thus, cannabinoids have an advantage of combining several neuroprotective mechanisms, something that is especially important in the treatment of neurodegenerative disorders in which neuronal damage is the consequence of a progressive consequence of different types of cytotoxic events (Fernandez-Ruiz et al. 2014).

The neuroprotective effects of THC are mediated by its action on both CB_1 and CB_2 receptors. Endogenous ligands for those receptors, endocannabinoids, produce neuroprotection by opposing the effects of stimuli that damage the brain (Pacher and Mechoulam 2011). Activation of the endocannabinoid system has been observed in some neurodegenerative conditions, including brain trauma in neonatal and adult rats (Hansen, Schmid, et al. 2001; Panikashvili et al. 2001), experimental Parkinsonism in rats (Gubellini et al. 2002), and kainate-induced excitotoxicity in mice (Marsicano et al. 2003). The upregulation of CB_1 receptors has been observed after experimental stroke (Jin, Mao, Goldsmith, and Greenberg 2000), after excitotoxic stimuli in neonatal rats (Hansen, Ikonomidou, Bittigau, Hansen, and Hansen 2001), and in the postmortem basal ganglia of Parkinson's Disease patients (Fernandez-Ruiz et al. 2014; Lastres-Becker et al. 2001). However, since CB_1 receptors are primarily located in neurons within the CNS, the response of CB_1 receptors is reduced in the neurons that degenerate in most neurodegenerative disorders. In cases in which upregulation of CB_1 receptors was found, the response appears to occur only on surviving neurons or in neurons other than those affected by the disease. CB_2 receptors, on the other hand, display a marked upregulation in all neurodegenerative disorders, including Alzheimer's Disease, Huntington's Disease, amyotrophic lateral sclerosis (ALS), and Parkinson's Disease (Fernandez-Ruiz et al. 2014). In the normal healthy brain, CB_2 receptors are located primarily on glial cells, including astrocytes and oligodendrocytes. Although there is some controversy in the literature, it is generally understood that they are absent from non-activated microglial cells. These CB_2 receptors upregulate in response to inflammatory, excitotoxic, infectious, traumatic, or oxidant insults occurring in most neurodegenerative disorders, and this upregulation is extremely intense in reactive microglial cells recruited to lesioned sites (Fernandez-Ruiz, Garcia, Sagredo, Gomez-Ruiz, and de Lago 2010; Fernandez-Ruiz et al. 2007).

Neuroprotective effects of CB_1 agonists

The most important neuroprotective property of CB_1 agonists is the normalization of glutamate homeostasis (Fernandez-Ruiz et al. 2010). The excitotoxicity process of excessive glutamate activity results in intracellular accumulation of calcium, which activates numerous destructive pathways, leading to cell swelling and death. The activation of CB_1 receptors opposes glutamatergic cytotoxic events. It reduces the excessive glutamate release on pre-synaptic neurons and reduces the excessive intracellular levels of calcium on post-synaptic neurons located on neurons containing NMDA receptors (Abood, Rizvi, Sallapudi, and McAllister 2001; Shen and Thayer 1998).

CB_1 agonists also increase the supply of blood to the injured brain, an effect particularly relevant to stroke. This effect is related to the reduction in the levels of some vasoconstrictor factors (Fernandez-Ruiz and Gonzales 2005). Brain damage during stroke or traumatic injuries is associated with release of endothelin-1. Endothelin-1 then produces vasoconstriction, thereby limiting the supply of blood to the injured area and aggravating damage to the brain. CB_1 agonists modulate vascular tone and therefore may provide neuroprotection by decreasing endothelin-1, which restores blood supply to the tissue (Mechoulam 2002). This effect is mediated by CB_1 receptors in brain microvasculature (Hillard 2000), but CB_2 receptors may also be involved.

Neuroprotective effects of CB_2 agonists

The primary neuroprotective action of CB_2 receptors is control of glial effects on neurons (Fernandez-Ruiz et al. 2010). Acute and chronic degenerative brain pathologies result in proliferation, recruitment, activation, and migration of both astrocytes and microglia, which can produce positive and negative effects, respectively. Astrogliosis has positive influences by generation of pro-survival and neurotropic factors (Allaman, Belanger, and Magistretti 2011), but activation of microglial cells into a reactive state has been associated with greater neuronal injury (Cunningham, Martinez-Cerdeno, and Noctor 2013). The classic detrimental effects of reactive microglial cells are exerted by generation of neurotoxic mediators, including tumor necrosis factor α (TNF-α), interleukin (IL)-1β, IL-6, eicosanoids, nitrous oxide, and reactive oxygen species that produce neuronal damage. Targeting CB_2 receptors, alone or in concert with CB_1 receptors, may improve neuronal homeostasis (Fernandez-Ruiz et al. 2014).

Cannabinoids as anti-oxidants

Cannabinoids also have neuroprotective effects that are not mediated by CB_1 or CB_2 receptors, among them the anti-oxidant properties of CBD and THC. CBD has a broad spectrum of potential therapeutic effects, including neuroprotective effects, but it does not bind well to CB_1 receptor or to CB_2 receptors. CBD protects against the brain damage produced by alterations in glutamate homeostasis (El-Remessy et al. 2003; Hampson et al. 2000), by oxidative stress, and by local inflammatory events, perhaps through its ability to inhibit endocannabinoid inactivation (Bisogno et al. 2001). CBD contains two hydroxyl groups that enable it to perform an important anti-oxidant activity: scavenging reactive oxygen species (Fernandez-Ruiz et al. 2014).

Cannabinoids and Multiple Sclerosis

Multiple sclerosis (MS) is an inflammatory disease of the brain and the spinal cord in which lymphocytic infiltration leads to damage of myelin and axons. Early in the course of the disease, inflammation is transient and remyelination occurs; therefore, patients usually recover from symptoms of neurological dysfunction. Over time, widespread microglial activation associated with chronic neurodegeneration occurs, accompanied by progressive disability. Currently available medications reduce the frequency of new episodes but do not reverse the course of the disease (Compston and Coles 2008). The number of cases of MS around the world has increased steadily, perhaps because of improved diagnosis; it now affects 2–3 million people (Kurtzke 1993). MS is three times as common in females as in males (Pryce and Baker 2014).

It is now known that axonal loss rather than myelin damage is the primary determinant of progressive disability in MS. In addition, doubling in the levels of glutamate is seen in the cerebrospinal fluid of MS patients undergoing an inflammatory episode (Stover et al. 1997), which suggests a loss of homeostatic control of neurotransmission. This elevation of glutamate may result from excessive signaling of excitatory circuits due to loss of inhibitory circuits. Cannabinoids may control these symptoms.

Cannabinoids and management of MS symptoms

Among the symptoms of MS that patients claim cannabis alleviates are bladder incontinence, tremor, and limb spasticity (Consroe, Musty, Rein, Tillery, and Pertwee 1997). Spasticity, one of the most commonly reported symptoms, may affect approximately 50 percent of patients. Current therapies for spasticity include the baclofen, tinazidine, and benzodiazepines (Paisley, Beard, Hunn, and Wight 2002). However, in a recent German study adverse negative side effects were reported in 92 percent of patients and poor efficacy in 88 percent of patients (Henze, Flachenecker, and Zettl 2013), and patients are seeking alternative medications. The pathophysiology of spasticity is not well understood, but it may reflect a loss of inhibitory circuitry in the spinal cord that allows excessive stimulation, which can result in excessive contractions of the muscles, sometimes even when the patient is at rest (Pryce and Baker 2014).

In mouse models of multiple sclerosis, CB_1 agonists inhibit spasticity and CB_1 antagonists worsen it (Pryce and Baker 2007). This pattern of findings suggests that endocannabinoids modulate spasticity in multiple sclerosis (Baker et al. 2000). Sativex (2.7 mg THC: 2.5 mg CBD/spray) alleviated hind-limb spasticity in the mouse model of MS (Hilliard et al. 2012). In addition, AEA levels are raised in the spinal cords and brains of mice showing hind-limb spasticity, but not in animals with equivalent neurodegeneration not displaying spasticity (Baker et al. 2001). Finally, administration of FAAH inhibitors (which elevate AEA) also reduces the level of spasticity in mice (Pryce and Baker 2014). Therefore, the pre-clinical data provide clear evidence that cannabinoids have therapeutic potential for MS.

A thorough systematic review of the efficacy and safety of medical marijuana in selected neurological disorders titled *Report of the Guide-line Development Subcommittee of the American Academy of Neurol-ogy* (Koppell et al. 2014) was recently published by the American Academy of Neurology. The review included studies conducted from 1948 to November of 2013 to address treatment of symptoms of multi-ple sclerosis, epilepsy, and other movement disorders. When the studies were graded according to the American Academy of Neurology's classifi-cation scheme for therapeutic articles, 34 of them met inclusion criteria and eight were rated as Class 1. The cannabinoid formulations used in those studies included oral cannabis extract, oral THC (Marinol and Cessamet), nabiximols (Sativex 2.7 mg THC: 2.5 mg CBD/spray) admin-istered by oromucosal spray, and smoked marijuana. The review did not differentiate between CBD and mixtures of THC and CBD mixtures among the oral cannabis formulations. The oral cannabis extract formu-lations included 100 mg CBD, 2.5 mg THC: 0.9 mg CBD, and 2.5 mg THC: 1.25 mg CBD (Cannador, IKF Berlin). The review concludes that oral cannabis extract is effective and that nabiximols and THC probably are effective for reducing patient-centered measures (self-reports of improvement) of MS spasticity, and both oral cannabis extract and THC may be effective for reducing both patient-centered and objective mea-sures after a year. For central pain or painful spasms associated with MS (excluding neuropathic pain), oral cannabis extract is effective and THC and nabiximols probably are effective. For MS bladder dysfunction, nabiximols, THC, and oral cannabis extract probably are effective for reducing the frequency of bladder voiding. For tremors, both THC and oral cannabis probably are not effective in reducing tremors, but nabiximols probably is effective. The risk of severe adverse effects of oral cannabis extract was only 1 percent. The comparative effectiveness of medical marijuana and other therapies for these indications is not known.

Nabiximols (Sativex) has been used in Canada, the UK and Spain to treat spasticity associated with MS, and in Canada it also has been used to treat neuropathic pain in MS and as an add-on treatment to strong opioid therapy in patients with advanced cancer. No significant abuse or diversion of nabiximols has been observed thus far, and there is no evi-dence of tolerance to the effects of nabiximols (Robson 2011). Over many months of treatment in a long-term safety study, Wade, Makela, House, Bateman, and Robson (2006) reported that the mean dose of nabiximols actually tended to decrease. And in two clinical trials, nabiximols treatment was abruptly stopped to assess potential withdrawal symptoms (Notcutt, Langford, Davies, Ratcliffe, and Potts 2012; Wade et al. 2006); neither group reported observing a withdrawal syndrome (Robson 2011).

A recent randomized double-blind crossover study compared the abuse liability of nabiximols with that of dronabinol in healthy male and female participants with a history of non-dependent but regular recreational use of marijuana (Schoedel et al. 2011). The pharmacokinetics of nabiximols and oral THC were similar, with onset after 1–3 hours. Participant received, in random order, single administrations (separated by a minimum of seven days) of placebo, nabiximols (2.7 mg THC: 2.5 mg CBD/spray—four sprays, eight sprays, and sixteen sprays), and dronabinol (20 and 40 mg). Standard measures of drug discrimination and drug liking, subjective experiences, and cognitive function were recorded for each exposure. Nabiximols did not produce significant adverse cognitive or psychomotor side effects and showed lower abuse potential than dronabinol at lower doses. However, both medications at the highest doses exhibited some abuse potential, defined as self-reported liking for a drug compared with placebo. Therefore, careful monitoring of abuse and of aberrant medication-related behaviors during clinical treatment with nabiximols is warranted.

Cannabinoids as neuroprotective therapy in MS

There is increasing evidence that elevated levels of glutamate are seen not only in MS patients, but also in animal models exhibiting MS spasticity accompanied by an increased level of expression of Group 1 metabotropic glutamate receptors and excitatory amino-acid transporters (Sulkowski, Dabrowska-Bouta, Kwiatkowska-Patzer, and Struzynska 2009). Modulation of the effects of elevated CNS glutamate shows disease amelioration in experimental studies and clinical studies (Pryce and Baker 2014). Endocannabinoids may have neuroprotective properties in neuroinflammatory disease by downregulating the release of glutamate. In a model of MS with CB_1 knockout mice, neuroinflammation resulted in accelerated accumulation of neurological deficits relative to wild-type mice (Jackson, Pryce, Diemel, and Baker 2005; Pryce et al. 2003). Likewise, administration of CB_1 agonists in wild-type mice can inhibit neurodegeneration due to neuroinflammation both in models of acute disease and in models of chronic disease. These findings suggest that cannabinoid therapy may have potential to slow neurodegeneration from MS and may be considered as an adjunct therapy to current disease-modifying therapies (Pryce and Baker 2014).

Only one clinical trial has investigated whether or not cannabinoid therapy may slow the neurodegeneration that causes the progression of MS (Zajicek et al. 2013). A total of 493 MS patients with primary or secondary progressive MS were recruited over a two-year period in the United Kingdom. A requirement for inclusion was that their walking was disrupted, but not prevented, by their MS. Participants were randomly assigned to receive THC capsules or placebo capsules over a period of three years. The first four weeks of the trial were used to establish the best-tolerated dose, which was then to be used in the remainder of the study period. The dose was gradually reduced to zero at the end of the treatment period. Despite the abundant pre-clinical experimental evidence suggesting that THC has a neuroprotective role in neurological diseases, the study found no evidence that THC affected the progression of MS as assessed either by neurological assessments by physicians or by participants' responses to questionnaire using a Multiple Sclerosis Impact Scale. There was some evidence, however, that in participants at the lower end of the disability scale THC had statistically significantly ($p < 0.01$) beneficial effects on the neurological assessments by physicians relative to participants given placebo. Since this represented a small percentage of the participants, further studies are necessary targeting patients at the lower end of the disability spectrum to determine if THC can protect against MS progression (Pryce and Baker 2014).

Cannabinoids for chronic neuropathic pain in MS

As was discussed in chapter 9, there is considerable evidence from experimental models of the efficacy of cannabinoids in reducing pain, and chronic pain is a frequent symptom of MS. Central neuropathic pain caused by a lesion or dysfunction of the CNS is a common symptom of MS, affecting between 17 percent and 52 percent of patients (Langford et al. 2013). Pain of this kind is difficult to treat. Nabiximols (Sativex, 2.7 mg THC: 2.7 mg CBD/spray) administered via sublingual spray is one potential treatment. A recent study showed beneficial effects of nabiximols relative to placebo (Rog, Nurmikko, Friede, and Young 2005). In phase II and phase III studies, nabiximols was found to have analgesic properties that were effective in relieving neuropathic pain. Those studies also showed that nabiximols was tolerated well and that it could improve sleep and quality of life. A subsequent two-year open-label follow-up study showed continued effectiveness of nabiximols spray and no evidence of tolerance in the 28 participants who completed the study (Rog, Nurmikko, and Young 2007). A meta-analysis (Iskedjian, Bereza, Gordon, Piwko, and Einarson 2007) of the effectiveness of nabiximols, dronabinol, and CBD in alleviating neuropathic and MS-related pain revealed statistically significant pain relief in these studies, concluding that some patients do not obtain relief but others respond very well.

A recent phase III clinical trial (Langford et al. 2013) assessed the efficacy of nabiximols spray for alleviating central neuropathic pain specifically due to MS. MS patients who did not respond to conventional pain therapies were treated with nabiximols or placebo as an add-on treatment. Phase A was a double-blind, randomized, placebo-controlled 14-week treatment period. Mean daily pain was scored by self-report. In phase B, all patients were treated with nabiximols for 14 weeks, followed by a 4-week randomized withdrawal phase during which patients received either nabiximols or placebo in a double-blind manner. The primary measure was the mean daily pain scores during the withdrawal period.

During the first 14 weeks (phase A), the self-reported pain did not differ significantly between the patients treated with nabiximols and the placebo controls. In phase B, during the randomized withdrawal period, the group that continued with the nabiximols treatment showed reduced pain scores, but the group changed to placebo did not. The results of this study were therefore equivocal with somewhat conflicting findings in the two phases, suggesting that further studies are warranted to evaluate the potential of nabiximols to treat chronic pain in MS patients.

Cannabinoids in treatment of MS: Summary and conclusions

Cannabis and medications derived from it have been confirmed to alleviate MS-related spasticity in both experimental and clinical settings, and nabiximols have been added to the list of medications in various countries. An issue that has not been resolved in many countries is the economic cost of the drug to a patient suffering from MS. As with all cannabis medications, efficacy will have to be balanced with the well-known side effects of cannabis caused by global stimulation of cannabinoid receptors in the brain rather than stimulation only in the regions needed to subside the symptoms—effects that many patients find undesirable. Pre-clinical experimental work with animal models has shown that boosting endocannabinoid levels by inactivation of FAAH or MAGL has some efficacy in treatment of spasticity (Baker et al. 2001; Baker,

Pryce, Jackson, Bolton, and Giovannoni 2012). Future medications that harvest the potential to boost natural endocannabinoid levels only in the regions of the brain where elevation of endocannabinoid levels is necessary to treat spasticity in MS may greatly improve the treatment of this disorder without the psychoactive side effects. These new experimental approaches may lead to the next generation of cannabinoid therapeutics for MS.

Cannabinoids and Alzheimer's Disease

Profound cognitive impairment is evident in older individuals with late-onset Alzheimer's Disease (AD), for which there is a lack of effective treatments. Recently, cannabinoids have been proposed as a treatment for this disorder (Ahmed, van der Marck, van den Elsen, and Olde Rikkert 2015). AD is characterized by a decline in cognitive and intellectual functions that interferes with daily living. "Late-onset AD" refers to AD diagnosed after the age of 60. The brains of patients with AD show accumulation of amyloid-β protein in extracellular senile plaques in various brain regions, but particularly in the hippocampus, in the cerebral prefrontal cortex, and in the amygdala. A second marker of AD is the presence of intracellular neurofibrillary tangles formed by hyperphosphorylated tau protein. These changes in brain function are believed to be produced by neuroinflammation and oxidated stress, but the actual cause is still unknown. The accumulation of senile plaques leads to chronic activation of microglial cells and astrocytes, which surround the plaques. This inflammatory response leads to the release of neurotoxic substances that produce neuronal death, ultimately resulting in cognitive deficits and behavioral changes.

Currently approved treatments for AD include the cholinesterase inhibitors donepezil, rivastigmine, and galantamine and the N-methyl-D-aspartate-receptor antagonist mematine. Each of these treatments acts on symptoms but does not have a profound disease-modifying effect. Because the neuropathology of AD involves multiple mechanisms, a treatment strategy focusing on multiple targets may be more beneficial than one that focuses on one target. Almost all patients with late-onset AD also develop neuropsychiatric symptoms, such as depression, anxiety, aggression, sleep disorders, and eating disorders. The currently approved treatments for AD have no effect on these neuropsychiatric symptoms (Ahmed et al. 2015).

Targeting the endocannabinoid system has been proposed as an approach to the treatment of AD. Numerous *in vitro* and *in vivo* studies have demonstrated the protective effects of cannabinoids against amyloid-β peptide and tau phosphorylation, the markers of AD. (See Ahmed et al. 2015 for a review.) Both AEA and 2-AG have been shown to reduce amyloid-β neurotoxicity by a CB_1-receptor mechanism (Milton 2002). Ramírez et al. (2005) demonstrated that senile plaques in AD patients express cannabinoid receptors CB_1 and CB_2, together with markers of microglial activation, and that CB_1 positive neurons are greatly reduced in areas of microglial activation. In addition, in a pre-clinical investigation they found that intracerebroventricular administration of a CB_1 agonist to rats prevented amyloid-β-induced microglial activation, cognitive impairment, and loss of neuronal markers. In a subsequent *in vitro* study, both CB_1 agonists and CB_2 agonists blocked amyloid-β-induced activation of cultured microglial cells and reduced neurotoxicity after amyloid-β addition to rat cortical cell cultures (Ramirez, Blazquez, Gomez del Pulgar, Guzman, and de Ceballos 2005). These results indicate that cannabinoid receptors are important in the pathology of AD and may be effective in preventing the neurogenerative process that occurs in Alzheimer's Disease.

Although pre-clinical and *in vitro* research suggest that agonists of CB_1 receptors may be promising treatments for reducing amyloid-β activation, the memory-impairing effects of such treatments discussed in chapter 5 may limit their usefulness. However, Chen et al. (2013) demonstrated that administration of a COX_2 inhibitor (e.g., ibuprofen) can prevent the memory-impairing effects of THC. Indeed, they show that daily injection of THC for four weeks reduced both amyloid-β and neurodegeneration in 5XFAD APP transgenic mice (a model of Alzheimer's Disease) whether or not the injections were accompanied by administration of a COX_2 inhibitor. Therefore, adjunct COX_2 inhibition did not affect this particular beneficial effect of THC. These findings are exciting because they suggest that COX_2 inhibitors may enhance the medical utility of marijuana by reducing the side effect of memory impairment (Chen et al. 2013).

The protective effects of cannabinoids have been shown not only with CB_1 agonists similar to THC but also with CBD. CBD has been proposed as an anti-oxidant neuroprotective agent because it inhibits *in vivo* amyloid-β plaque formation and decreases reactive oxygen species production. CBD also inhibits the hyperphosporylation of tau protein *in vitro* in a neuronal cell line. CBD actually reduces the formation of amyloid plaques and neurofibrillary tangles. Recent work with a combination of equidoses (0.75 mg/kg daily for five weeks) of THC and CBD in a mouse model of AD resulted in reduced cognitive impairment and reduced amyloid plaques (Aso, Sanchez-Pla, Vegas-Lozano, Maldonado, and Ferrer 2015).

There have been very few human clinical studies with THC as a treatment for symptoms of dementia, and none with CBD. Among four clinical trials of treatment of dementia symptoms (with only 60 participants in all), all participants were treated with the synthetic THC dronabinol. The primary measure for each of these studies was not cognition; however, cognitive measures were taken as secondary measures, and in all but one dronabinol improved cognition (van den Elsen et al. 2015) without severe side effects or medication discontinuation (Ahmed et al. 2015). In a promising though limited recent study, Shelef et al. (2016) evaluated the efficacy and safety of medical cannabis oil containing THC (2.5 mg twice a day) as an add-on to pharmacotherapy (with eight anti-psychotic medicines for neuropsychiatric symptoms and four acetycholinesterase inhibitors) in relieving behavioral and psychological symptoms of dementia in eleven AD patients. Ten of the eleven patients recruited to a four-week open-label prospective trial completed the trial. Those ten patients showed significant reductions in severity of clinical symptoms and in neuropsychiatric symptoms (delusions, agitation/aggression, irritability, apathy, sleep, caregiver distress). Therefore, adding medical cannabis oil to a patient's pharmacotherapy may be a safe and promising treatment for AD.

Cannabinoids and Ischemia, Brain Trauma, and Spinal Injury

Acute neurodegeneration due to ischemic stroke, perinatal hypoxia-ischemia, spinal-cord injury, and brain trauma can be a major cause of permanent disability. The primary injury leads to secondary damage, including neuroinflammation, glutamatergic excitotoxicity, calcium influx, caspase activation, vasoconstriction, and oxidative stress (Moskowitz 2010). Because there are multiple pathways to secondary damage, cannabinoids have been studied in pre-clinical models over the past 15 years. *In vivo* treatment with THC and with CBD reduced lesion expansion and neurological deficits in animal models of acute neurodegeneration, including local or global ischemia, contusive spinal-cord injury, closed head injury, and newborn hypoxia ischemia (Fernandez-Ruiz et al. 2014). Both THC and CBD have effectively preserved neuronal survival in these studies. Most studies indicate that the effect of THC is mediated by CB_1 receptors in these beneficial effects. Indeed, CB_1 knockout mice showed increased mortality from permanent focal ischemia and larger infarcts after transient focal cerebral ischemia relative to wild-type mice (Parmentier-Batteur, Jin, Mao, Xie, and Greenberg 2002). CBD's effects appear to be mediated by its anti-oxidant properties (Fernandez-Ruiz et al. 2013). Although the pre-clinical evidence is promising, there have been no clinical trials. Novel clinical studies of THC, CBD, and Sativex with stroke patients and brain-trauma patients are warranted (Fernandez-Ruiz et al. 2014).

Cannabinoids and Huntington's Disease

Huntington's Disease (HD) is an inherited progressive neurodegenerative disease. A mutation in the huntingtin gene (IT15) affects) the medium spiny GABAergic striatal neurons associated with the characteristic "choreic" movements that appear in the early stages of the disease and the glutamatergic neurons that project from the cortex to the striatum and are associated with the cognitive dysfunctions and the psychiatric symptoms that appear in HD.

The disease is typically treated with neuroleptic drugs. The only drug currently approved (tetrabenazine, an inhibitor of the monoamine vesicular transporter) has only small effects on symptoms in HD patients (Chen, Ondo, Dashtipour, and Swope 2012). There are no approved therapies for the progression of HD.

Cannabinoids have been proposed as candidates for a neuroprotective therapy in HD, which may involve CB_1 and CB_2 receptors as well as receptor-independent mechanisms (Sagredo, Ramos, Decio, Mechoulam, and Fernandez-Ruiz 2007). Blazquez et al. (2011) demonstrated that defects in CB_1-receptor signaling in the basal ganglia may initiate excitotoxicity, so early stimulation of these striatal receptors might reduce the progression of striatal degeneration (Blazquez et al. 2011). CB_2 receptors are upregulated in astrocytes and reactive microglial cells activated by the lesioned striatum (Bouchard et al. 2012; Palazuelos et al. 2009); thus, CB_2 agonists protect striatal projection cells from death by reducing the toxicity of reactive microglial cells and enhancing the trophic support by astrocytes (Fernandez-Ruiz et al. 2014). Aside from the CB_1-mediated and CB_2-mediated actions of phytocannabinoids on HD, there is also evidence that CBD may be neuroprotective in this disease, by acting as an anti-oxidant. CB_1 activation reduces excitotoxicity, CB_2 activation reduces inflammation, and CBD decreases oxidative injury in HD.

All the evidence to date for the potential of cannabinoids to protect against the progression and/or symptoms of HD is based on pre-clinical animal models. Early clinical trials with CBD were aimed at treating symptoms such as chorea, but were without success (Consroe et al. 1991). The results of clinical studies with nabilone have reported some improvement in motor symptoms (Curtis, Mitchell, Patel, Ives, and Rickards 2009; Curtis and Rickards 2006); however, nabilone was also reported to have worsened chorea in a single-patient case study (Muller-Vahl, Schneider, and Emrich 1999). More recently, the first clinical trial that actually monitored the progression of HD (rather than the effect on symptom relief) was conducted with Sativex (2.7 mg THC: 2.5 mg CBD/ spray). Unfortunately, there was no evidence of slower disease progression, but the clinical trial demonstrated that the medicine was safe and was well tolerated by HD patients (Fernandez-Ruiz et al. 2014).

Cannabinoids and Parkinson's Disease

Parkinson's Disease (PD) results from progressive degeneration of DA neurons of the substantia nigra, which may be triggered by genetic risk factors in combination with unknown environmental factors (Fernandez-Ruiz et al. 2014). The symptoms of PD, which include bradykinesia, resting tremor, rigidity, and postural disturbances, can be treated with L-dopa, dopamine agonists and monoamine oxidase inhibitors (all aimed at replacing lost dopamine). Current treatments also include deep brain stimulation and replacement of degenerated neurons with dopamine-producing cells.

Phytocannabinoids have been proposed as a promising therapy (Garcia-Arencibia, Garcia, and Fernandez-Ruiz 2009). In particular, CBD, THC, and tetrahydrocannabivarin (THCV) have been reported to be neuroprotective in rats lesioned with 6-hydroxydopamine (which selectively destroys DA neurons) by virtue of their non-cannabinoid anti-oxidative properties (Garcia-Arencibia et al. 2009; Lastres-Becker, Molina-Holgado, Ramos, Mechoulam, and Fernandez-Ruiz 2005). In addition, activation of CB_1 receptors may be involved in the neuroprotective effects of THC; mice lacking this receptor were more sensitive to the DA-damaging effects of the dopamine neuron neurotoxin, 6-hydroxydopamine (Perez-Rial et al. 2011). Since THCV acts as a CB_1 antagonist, its anti-oxidant properties on PD may be a better alternative, because other CB_1 antagonists reduce bradykinesia in animal models of PD (Fernandez-Ruiz 2009). THC's anti-inflammatory effects on CB_2 receptors may benefit PD; indeed, the selective CB_2 agonist HU-308 preserved DA neurons in animal models of PD (Ternianov et al. 2012), and CB_2-receptor-deficient mice were more vulnerable to DA damage than wild-type mice in these models (Garcia et al. 2011; Price et al. 2009). However, to date no clinical trials of cannabinoids have been reported in PD patients with either CB_2 agonists or THCV.

Cannabinoids and Amyotrophic Lateral Sclerosis

Amyotrophic lateral sclerosis (ALS), known as motor neuron disease, is a neurodegenerative disease involving selective injury and death of motor neurons in the spinal cord, the brainstem, and the motor cortex. The etiology of ALS is currently not known, although it is known that about 10 percent of cases are genetic (Ahmed and Wicklund 2011). The only treatment for ALS is Riluzol, an anti-glutamatergic agent that acts by blocking voltage-dependent sodium channels on motor neurons and thereby reducing their activity (Cheah, Vucic, Krishnan, and Kiernan 2010). Recent work suggests that THC delayed motor impairment and increased survival in a mouse model of ALS (Raman et al. 2004). FAAH knockout mice with permanent elevation of AEA also showed a delayed onset of the disease (Bilsland et al. 2006). There have been no clinical trials of cannabinoid drugs with human ALS patients.

Conclusion

The best evidence for the effectiveness of cannabis as a medicine comes from its use to treat painful spasticity in MS patients. Indeed, Sativex is currently available as a prescribed treatment for this condition in Canada. Evidence from human clinical trials in patients with other neurodegenerative disorders, including AD, PD, HD, and ALS, is less clear; however, the *in vivo* and *in vitro* pre-clinical evidence has demonstrated that the endocannabinoid system is involved in symptomology.

12

Where Do We Go from Here?

These are interesting times. As of this writing, many countries around the world, and 23 of the United States, have legalized medical marijuana, and four US states and the District of Columbia have legalized recreational marijuana. The premier of Canada, Justin Trudeau, has indicated his intention to legalize marijuana in that country. It seems that cannabis stands to become a legal drug, like alcohol and tobacco. For that reason, it is important to understand what consequences, both positive and negative, will occur if cannabis is legalized and more and more people are exposed to it.

In the US the estimated current yearly market for marijuana is $5.7 billion; four years from now it is expected to increase to $24 billion as a consequence of legalization (source: N. Volkow, "Marijuana and Cannabinoids: A Neuroscience Research Summit—Day 1, March 22, 2016," videocast.nih.gov). What can we do to minimize adverse effects?

We are entering a new era in understanding the risks and benefits of marijuana. The benefits of marijuana to appetite, pain, anxiety, nausea, epilepsy, and disorders of the nervous system have been observed throughout history, but now we are beginning to understand the mechanisms for these effects. Clearly, cannabinoids act on neural circuits, and dysfunction within these circuits is the basis of neurological disorders for which investigators are trying to find new treatments. Clinical observations are important, but in order to protect public health it is important to understand how and why such effects occur. For instance, CB_1 agonists have been reported to be both anti-convulsant and pro-convulsant in epilepsy. It is known that there are many kinds of epilepsy. Some epileptics might benefit, but others might experience harm. How can we predict the outcome?

The media have focused on the potential link between cannabis and psychosis, but dependence on cannabis use is a much more common adverse effect. Concern has been raised about cognitive and brain structural changes produced by cannabis, but their causality and longevity have been greatly contested. Identifying why some individuals are more vulnerable than others to the adverse effects of cannabis is of utmost importance for public health (Curran et al. 2016).

We know that cannabinoids act on CB_1 and CB_2 receptors, but they also act on TRPV1 receptors and on GPR55 receptors and have non-receptor effects on microglia, on cytokines, and on reactive oxygen species. All these effects are dependent on dose; often a low dose produces the opposite effect of a high dose. Long-term and short-term effects must be delineated. There is good evidence that marijuana reduces painful spasms, both static ones and dynamic ones triggered by noise or anxiety, in patients with multiple sclerosis. In studies of such treatment, what outcome measure are we trying to obtain, and what measure is most important? For instance, in one important review (Koppell et al. 2014), MS patients' perception of spasticity was decreased by the drug, but the objective measure made by the physician indicated no difference.

What Next for Medical Marijuana?

There is good, clear evidence that medical marijuana alleviates pain, especially chronic neuropathic pain. In view of the current rate of deaths due to overdoses of prescription opiates in the United States, there is a desperate need for alternatives to such opiates. The synergism evident in the literature on pre-clinical trials with opiates holds out hope that patients may be able to greatly reduce the doses of opiates they use to control pain, which may reduce the risk of addiction and that of death by overdose. One study (Abrams et al. 2011) demonstrated that vaporized marijuana can augment the analgesic effects of opioids; however, that study did not include double-blind placebo controls. Combined low doses of opiates and marijuana may well advance the treatment of chronic pain.

Pain patients report anecdotally a preference for smoked or vaporized marijuana rather than orally administered THC (nabilone or marinol). This probably is a result of faster onset and shorter duration of action, allowing titration of the dose to obtain the desired effect. However, this relative effectiveness has not been directly assessed in a double-blind placebo comparison. Ben Wilsey is currently conducting such a trial. (See B. Wilsey, "Smoked and vaporized cannabis in neuropathic pain," http://videocast.nih.gov Marijuana and Cannabinoids: A Neuroscience Research Summit—Day 1, March 22, 2016.)

In view of the psychoactive effects of THC, it is difficult to run a truly double-blind study using an inactive placebo as a comparison. One possibility is the use of an active placebo, such as the benzodiazepine lorazepam, which may not be indicated for the condition under study, such as chronic pain. If medical marijuana actually relieves pain on the visual analog scale that is typically used, the marijuana group should report less pain than the lorazepam group without being able to guess group assignment. Wilsey's group is currently comparing the pain-reducing effectiveness of active and inactive oral administration of THC against that of active and inactive vaporized administration.

It is important to understand the effects of the different cannabinoids in marijuana to understand potential therapeutic treatments. In particular, there is a desperate need for data on the most efficacious ratios of THC to CBD in the marijuana used to treat various conditions. To date, no clinical trials have looked at the efficacy of different ratios of THC to CBD in smoked or vaporized marijuana for treatment of pain or of any other condition, nor have any studies evaluated the potential of smoked or vaporized cannabis with high CBD to treat pain or any other condition. In fact, despite several small-scale studies, the only broad-based human clinical trials with CBD have been based on the use of sublingual spray (Epidiolex) to treat childhood epilepsy and on oral administration to treat psychosis (Leweke et al. 2012). The promising initial results using Epidiolex and perhaps high-CBD cannabis in treating Dravet Syndrome are encouraging. It will be also be necessary to understand how the beneficial medicinal effects of cannabis may be achieved without causing harm. Surveys indicate that recreational use of marijuana is accompanied by use of alcohol (Midanik, Tam, and Weisner 2007), yet very little is understood about the co-morbidity consequences of their interaction in the system. Does one alter the metabolism or the neurobiological effects of the other? Almost all we know about these drugs is about the individual drugs, but we know little about what happens when they are combined. This is especially important in relation to driving.

What Next for Endocannabinoids?

Of course the science behind cannabis is based on the science behind our own endogenous cannabinoid system. The endocannabinoids AEA and 2-AG are the first lipid-based neurotransmitters known to be produced on demand and then rapidly destroyed. They are functionally different, but in subtle ways. Future research will bring a better understanding of the separate and distinct roles of 2-AG and AEA. It is known that 2-AG is more broadly expressed than AEA and that it is tonically released as a point-to-point retrograde messenger to regulate a variety of neural plasticity processes. AEA, more of a modulatory transmitter, is involved in stress responsivity; it modulates stress in the central nervous system and pain in peripheral nervous system. Animals genetically modified to lack FAAH are relatively normal, because the resultant elevation in AEA is not revealed in their overall behavior, but they are less responsive to painful stimuli (Lichtman 2004). In order to see a change in their behavior, we must expose them to a stressful environment.

Instead of a single ligand acting at multiple receptors (as serotonin and dopamine do), the endocannabinoid system consists of different ligands acting at the same receptors. Although FAAH inhibitors and MAGL inhibitors produce few psychoactive effects, dual FAAH/MAGL inhibition has been shown to produce psychoactive behavioral effects more like those caused by delivery of a global CB_1 agonist (Wise et al. 2012). Future research will determine how 2-AG and AEA interact differently at receptors to produce differential effects. Do they compete for the same receptor, or do they act differently at the receptor to stimulate different signaling pathways (e.g., β arrestin/ERK; see Laprairie, Bagher, Kelly, and Denovan-Wright 2016)? Are they primarily localized in different regions of the brain? How do these two lipids interact with other lipid signals in the brain, such as prostaglandins? Future work may identify the means by which these endocannabinoids are transported across the cell membrane, and new therapies may be developed to prevent their re-uptake (Fu et al. 2012; O'Brien et al. 2013).

How do newly discovered negative endogenous allosteric modulators of the CB_1 receptor (e.g., pregnenolone) modify the effects of endocannabinoids and THC in cannabis to produce therapeutic effects? CBD has recently been reported to act *in vivo* as a negative allosteric modulator of the CB_1 receptor (Laprairie, Kulkarni, et al. 2016); does this account for the potential of CBD to reverse some of the effects of THC? CB_1 receptors have been discovered to occur on astrocytes (Han et al. 2012) and mitochondria; what is their function at those sites? Are CB_2 receptors found on neurons, not just on activated microglia at times of inflammation? We will learn more about the role of CB_2 receptors when better probes are developed for these receptors.

How can we use endocannabinoid signals for therapy? If we block the biological transformation of endocannabinoids by FAAH inhibition or MAGL inhibition, we can enhance the system's intrinsic regulatory functions with greater selectivity and safety than with global CB_1 agonists. Inhibition of FAAH is very promising in pre-clinical trials. It is clear that in animal models FAAH inhibition reduces anxiety, reduces depression, enhances social behavior, reduces nicotine addiction, and reduces nausea, all with very mild side effects. MAGL inhibition has been shown to reduce seizures, pain, nausea, anxiety, and withdrawal from morphine (Wills et al. 2016) or nicotine (Trigo and LeFoll 2015) in pre-clinical animal models, with relatively few side effects.

We know a lot about the safety of FAAH inhibitors in humans. Seven different FAAH inhibitors have been evaluated in eighteen phase I and phase II clinical trials involving more than 1,000 human subjects, with no severe side effects reported. (See D. Piomelli, "Therapeutic potential of the cannabinoid system," Marijuana and Cannabinoids: A Neuroscience Research Summit—Day 1, March 22, 2016, videocast.nih.gov.) However, a recent clinical trial conducted in France with a putative FAAH inhibitor manufactured by the Portuguese company Bial resulted in one death and five hospitalizations of paid volunteers. The drug was designed to act on the human endocannabinoid system as a potential painkiller and treatment for anxiety. Unfortunately, the details of the safety and efficacy of the compound have not been released. In a press release, Daniele Piomelli, editor-in chief of the new journal *Cannabis and Cannabinoid Research*, stated: "Several structurally different FAAH inhibitors have been previously tested for human safety in rigorous Phase 1 clinical trials. These include compounds from Sanofi, Pfizer, Merck, Johnson and Johnson, and others. All these FAAH inhibitors were shown to be safe." (source: press release from Mary Ann Liebert, Inc., New Rochelle, New York, January 19, 2016) Insofar as multiple FAAH inhibitors have been shown to be safe in humans, it is not likely that the toxicity of the Bial compound is due to its interaction with FAAH. "It is more probable," Piomelli continued, "that the Bial compound interacts with another as yet unknown protein that is responsible for the observed toxicity, or that a toxic impurity was present in the test drug. Of course, while we can tentatively exclude a class effect at this point, we cannot pinpoint which other target might be responsible for the toxicity of the Bial compound." It is unfortunate that the details of this study have not been released to the public, because this tragedy is likely to influence the acceptance of new FAAH inhibitors for testing in human clinical trials. Yet the pre-clinical evidence provides clear evidence of their potential in treating anxiety, depression, PTSD, pain, and nausea.

The value of pre-clinical animal data as a guide to the development of new treatments is evident in the story about rimonabant. Pre-clinical trials not only demonstrated its value in treating obesity and metabolic disorders but also revealed the psychiatric side effects of anxiety/depression, probably because of antagonism of AEA (stress responsivity regulator); therefore, we should trust these models when deciding how to proceed. One promising avenue for the treatment of obesity and metabolic disorders is the development of new compounds that bypass the central effects of CB_1 agonism or antagonism by not crossing the blood-brain barrier. Considerable pre-clinical research argues for the effectiveness of CB_1 antagonists in treating metabolic disorders without entering the brain. The development of selective CB_2-receptor agonists without the psychoactive effects of CB_1-receptor agonists is also likely to have great benefits for the treatment of pain and inflammation.

The Special Case of Palliative Care

The goals of palliative care include relief of pain nausea and enhancement of the quality of life (World Health Organization 2012). Medical use of cannabis is gaining some ground in palliative-care settings in which the focus is on individual choice, patient autonomy, empowerment, comfort, and quality of life. Careful assessment is required before administering cannabinoid drugs because certain patient populations may also be more sensitive to experiencing adverse psychotropic, cognitive, or psychiatric effects (Fine 2012); however, since the patients are near the end of life, it is important to ensure a high quality of life and to minimize pain and suffering.

Clearly, the evidence suggests that cannabis and cannabinoid drugs (dronabinol, nabilone, nabiximols) may be useful in alleviating a wide variety of symptoms often encountered in palliative-care settings, such as nausea and vomiting associated with chemotherapy, anorexia/cachexia, severe intractable pain, severe depressed mood, anxiety, and insomnia (Abrams and Guzman 2015). The use of cannabinoids for palliative care may make it possible to reduce the number of medications used by patients (Sutton and Daeninck 2006).

Recent observational studies conducted in palliative-care facilities in Israel have provided evidence for the potential of cannabis use to manage symptoms. More than 100 cannabis-using patients in a cancer palliative-care setting reported significant improvement in a variety of symptoms related to cancer and to anti-cancer treatment, including nausea, vomiting, mood disorders, fatigue, weight loss, anorexia, constipation, sexual dysfunction, sleep disorders, itching, and pain (Bar-Sela et al. 2013). Forty-three percent of the patients using pain medications and 33 percent of those using anti-depressants reported reductions in doses. No significant side effects except for memory lessening in patients with prolonged cannabis use were reported. Another observational study (Waissengrin, Urban, Leshem, Garty, and Wolf 2015) of the patterns of cannabis use among adult Israeli advanced cancer patients revealed that, of 17,000 cancer patients, 279 were authorized to use cannabis for medical purposes. The median age of the 279 patients was 60 years. Some had lung cancer (18 percent), some ovarian cancer (12 percent), some breast cancer (10 percent), some colon cancer (9 percent), and some pancreatic cancer (7.5 percent), and 84 percent of them had metastatic disease. Pain (76 percent) was the most common indication for prescription; other common indications were anorexia (56 percent), general weakness (52 percent), and nausea (41 percent). Improvement was reported for pain and general well-being (70 percent), appetite (60 percent), nausea and vomiting (50 percent), and anxiety (44 percent). Indeed, 83 percent rated the overall efficacy of cannabis as high. The most common route of administration was smoking (90 percent). Most patients (62 percent) reported no adverse effects, although 20 percent reported fatigue, 19 percent reported dizziness, and 6 percent reported delusions (Waissengrin et al. 2015). Observational studies clearly support the use of cannabis in treating palliative-care patients at the end of their lives.

Cannabinoid Designer Drugs

Cannabinoid designer drugs are synthesized compounds that function as agonists at CB_1 receptors and are used to produce intoxication similar to that produced by marijuana. Most of these drugs result from slight modifications to previously described chemical structures that were initially synthesized for research purposes by John Huffman (e.g., JWH-018) and Raphael Mechoulam (HU-210). The first of these compounds appeared around 2004. By 2008 there were many more.

Cannabinoid designer drugs now are sold under dozens of brand names, such as Spice and K2. They can be found on the Internet, in head shops, and in convenience stores. Legal controls are circumvented by rapid substitution of similar chemical structures not yet controlled. As one synthetic cannabinoid is banned, suppliers rapidly substitute an analogue to evade targeted detection. The popularity of synthetic cannabinoids has grown rapidly. One reason for their popularity may be that they usually aren't detected by standard urinary screening for drugs conducted in workplaces.

After cannabinol was chemically isolated (in the 1940s), a variety of novel analogues were synthesized and tested, and some extremely potent and long-acting dimethyheptylpryan (DMHP) analogues were discovered (Adams, Harfenist, and Loewe 1949). Those analogues turned out to be quite similar in structure to the yet-to-be-discovered principal psychoactive compound in marijuana, THC (Gaoni and Mechoulam 1964a). Indeed, it was these scientific advancements that led to the development of nabilone, the first synthetic cannabinoid to be approved for oral administration by the FDA for the treatment of nausea and vomiting in chemotherapy patients (It was approved in the 1970s.) Scientists' continuing development of synthetic variants of THC as research tools eventually led to the discovery that THC acts at G-protein-coupled receptors (GPCR)—that is, CB_1 and CB_2 receptors (Howlett, Johnson, Melvin, and Milne 1988). As these compounds were discovered, the information was made publicly available through publications or patents; it resulted in great advances in our understanding of the endocannabinoid system and in potential therapeutic options. However, the same information can also be used for clandestine purposes.

In 2004 cannabinoid designer drugs began to appear—first in Europe, where they were marketed as "herbal products" or "incense" to avoid detection by law enforcement. They quickly spread around the world. Analysis revealed that they were adulterated with synthetic cannabinoid agonists. By 2010 herbal products were sold openly in stores and on the Internet. The structures of the molecules most commonly found in these products were available in the published literature. Subsequent analysis of these compounds has revealed that they have excellent affinity for the CB_1 receptor—in fact, better affinity than THC, which only acts as a partial agonist (Thomas, Wiley, Pollard, and Grabenauer 2014). This may account for the anecdotal reports that the high they produce is more intense.

Systematic studies of the pharmacokinetics of the many synthetic cannabinoids have not been performed. However, it is known from existing data that they are lipophilic and therefore are absorbed very rapidly. Distribution to the brain of JWH-018 (a major constituent of synthetic cannabinoids) occurs within 60 minutes of inhalation and acts on CB_1 receptors. The majority of pharmacokinetic work has focused on urinary metabolites that can be used as forensic markers. Similar to THC, the first step in metabolism of synthetic cannabinoids is in the liver by cytochrome P450. Whereas THC has one major metabolite (11-OH-THC), the metabolism of synthetic cannabinoids follow several pathways, resulting in several metabolites, some of which are psychoactive. Since the metabolites of the various compounds differ, they are difficult to detect with standard urine testing.

Smokers of the drug called Spice typically report a cannabis-like high. However, it must be remembered that synthetic cannabinoids produce an effect similar to that of THC (but often more potent), but without the mitigating effects of CBD (potentially reducing anxiety, psychotic reactions, seizure potential, and memory deficits, as discussed in previous chapters). The side effects that have been reported—including agitation, anxiety, tachycardia, confusion, memory deficits, seizures, and psychotic reactions—may be reduced by co-administration of CBD. The most common complaint in surveys is tachycardia, cited in 40 percent of the 153 single-agent exposures reported to the US National Poison Data System in the first nine months of 2010 (Hoyte et al. 2012). The incidence of psychotic symptoms (9.4 percent) is also a concern. There is little information on long-term use of such drugs. There have been reports of deaths (of individuals from 15 to 35 years of age) linked to the use of synthetic cannabinoids in Australia (Gerostamoulos, Drummer, and Woodford 2015) and in the United States (Behonick et al. 2014); however, in each of these cases the cause of death was not ascertained even after a full autopsy and ancillary investigations. Sudden cardiac deaths have been reported in patients with a history of coronary heart disease; the side effect of tachycardia in one-third of users and that fact that most users of such drugs use alcohol with them may bear on this.

In sum, anecdotal data from hospital emergency rooms and poison centers suggest that synthetic cannabinoids are more dangerous than marijuana and might even be lethal in some cases. Legal controls, where they exist, are of limited effectiveness for three reasons: producers can replace banned chemical structures within weeks, Internet vendors are highly resourceful, and the demand is there. Detection is very difficult because the metabolites differ from THC (Thomas et al. 2014).

Other Potential Risks Associated with Recreational Marijuana

Throughout the book some of the potential risks of recreational marijuana have been discussed. They include neurocognitive dysfunction and increased risk of psychosis in vulnerable individuals, both effects being most problematic in adolescence. In addition, there are potential effects on the fetus or the newborn child when cannabis is used by a pregnant or a nursing woman, potential effects on the respiratory tract when cannabis is smoked, and, of course, potential effects on driving.

Pre-natal effects

Pre-natal exposure to cannabis through maternal use has been associated with neurodevelopmental differences in neonates and subsequent developmental changes in children (Jaques et al. 2014). The endocannabinoid system is critical for brain development; therefore, pre-natal activation of CB_1 receptors by cannabis may lead to neuroanatomical and behavioral changes in offspring (Alpar, Di Marzo, and Harkany 2016). Fetal growth may be affected by gestational cannabis exposure, but the dose-response relationship has not been identified (Jaques et al. 2014). There is pre-clinical rodent evidence that the suckling reflex is shaped by CB_1-receptor signaling pathways (Fride et al. 2001), and that disruption of those pathways by CB_1 antagonists can produce death in rodents. There is no evidence that pre-natal cannabis exposure increases the risk of subsequent development of schizophrenia or epilepsy (Alpar et al. 2016). However, in view of the currently available evidence of other effects, cannabis should not be used by pregnant or lactating women.

Lung cancer and respiratory disorders

Many of the potential carcinogens present in tobacco smoke are also present in cannabis smoke, yet a clear relationship between lung cancer and cannabis smoking has not been established when tobacco use has been controlled for (Joshi, Joshi, and Bartter 2014; Zhang et al. 2015). The evidence that cannabinoids may have anti-cancer activity through anti-metastatic and anti-angiogenic mechanisms (Cridge and Rosengren 2013; Ramer and Hinz 2016) is beyond the scope of this book.

The respiratory symptoms produced by smoking cannabis may be reduced by the use of vaporizers. Gieringer et al. (2004) found that a Volcano vaporizer's delivery of THC was similar to that of a cannabis cigarette but reduced the amount of carcinogens substantially. The rate of respiratory symptoms (bronchitis wheeze, breathlessness) among 150 persons who used only vaporizers and not cannabis cigarettes was 40 percent of that reported by cannabis smokers after controlling for cigarette smoking, duration of use, and amount typically used (Gieringer, St Laurent, and Goodrich 2004). Subsequently, data from a large Internet sample confirmed that use of a vaporizer predicted fewer respiratory symptoms even when age, sex, cigarette smoking, and amount of cannabis used were taken into account; however, since this was not a randomly assigned variable, there is a potential that cannabis users who choose to use a vaporizer engage in other behaviors to reduce respiratory diseases (Earleywine and Barnwell 2007).

Driving

Driving while intoxicated either by cannabis alone or by cannabis and other intoxicants appears to increase the risk of being in a motor-vehicle accident (Asbridge, Hayden, and Cartwright 2012; Hartman and Huestis 2013; Li et al. 2012). THC impairs perception, psychomotor performance, cognitive functions, and affective functions, all of which may contribute to a driver's increased risk of causing a traffic accident. After alcohol, cannabis and benzodiazepines are the drugs most often found in impaired drivers and in drivers involved in accidents. Experimental data show that people attempt to compensate by driving more slowly after smoking cannabis, but that control deteriorates with increasing task complexity. Cannabis smoking impairs cognitive function and increases lane weaving. Critical-tracking tests, reaction times, divided-attention tasks, and lane-position variability all show cannabis-induced impairment (Hartman and Huestis 2013).

In a large controlled case study conducted in the United States, the presence of THC or its metabolites in blood or urine was associated with an increase in potentially unsafe driving actions that increased the likelihood of contributing to an accident by approximately 30 percent (Bedard, Dubois, and Weaver 2007). Although initial reports indicated that cannabis use increased the risk of traffic accidents by 100 percent (Asbridge et al. 2012 ; Li et al. 2012), Rogeberg and Elvik (2016) reported in a more recent large meta-analysis of nine studies across five countries with an estimated sample count of 239,739 that cannabis use increased the risk of traffic accidents by 30 percent, bringing it into agreement with the ratio reported by Bedard et al. This later meta-analysis controlled for methodological shortcomings of previous studies, including better control for alcohol intoxication. Nevertheless, there remains a higher risk of accidents under the influence of cannabis, and the evidence indicates that recreational cannabis users should not drive while intoxicated. However, it is difficult to detect THC. Current detection screens require blood samples. THC concentrations in blood decrease rapidly after the end of smoking (within 30 minutes), yet typically blood samples are not taken for 1–4 hours after an incident. At present there are no simple and accurate roadside tests (Hartman and Huestis 2013).

How Does Variation among Strains of Cannabis Influence Desired and Undesired Effects?

One variable that recent research suggests may mediate the association between cannabis and addiction, psychosis, and cognitive impairment is the type and the level of cannabinoids in the cannabis that the individual smokes. (See Morgan et al. 2010.)

Connoisseurs of cannabis discuss the pros and cons of different varieties much as wine connoisseurs discuss grapes. What differentiates these different types of cannabis biochemically is probably the combinations and quantities of the more than 100 cannabinoids and considerable other compounds (e.g., terpenes) in any strain. We know the most about THC/CBD.

The ratio of THC to CBD in cannabis varies greatly. Levels of CBD vary from 0 to 40 percent. Higher levels of THC and negligible CBD have been found in varieties grown hydroponically; such strains are increasingly dominating the market. Although data on the THC content of cannabis have become increasingly available in the recent years, the levels of other cannabinoids are seldom measured. In one study conducted in the United States, ElSohly et al. (2016) analyzed more than 30,000 samples confiscated between 1994 and 2004 and found that the THC content of cannabis had increased by a factor of 3 in the resin and by a factor of 2 in the leaf, but that the CBD content remained the same in both leaf and resin during that period. Therefore, the ratio of THC to CBD increased during that period. However, we do not know how the increased THC content affects individual intake—that is, people probably titrate the dose. Freeman et al. (2014) measured concentrations of THC and CBD in cannabis provided by participants, then had each participant roll a joint and smoke it in front of the researcher. THC concentrations were found to be related negatively to the amount of cannabis used. When using their own cannabis in a naturalistic setting, people titrate the amount they roll in cigarettes according to the concentration of THC, but not according to the concentration of CBD (Freeman et al. 2014).

In the pre-clinical animal literature, CBD has been reported to enhance some of THC's effects and to reduce others. In the case of nausea (Rock, Limebeer, and Parker 2015) and in the case of pain (Varvel et al. 2006), synergy between CBD and THC produces a dramatically improved therapeutic effect when sub-threshold doses are combined. On the other hand, CBD reduces THC's memory-impairing effects (Morgan, Schafer, et al. 2010), its anxiety effects (Zuardi et al. 1981), its psychoactive effects, and its psychotic side effects (Bhattacharyya et al. 2010; Fusar-Poli et al. 2009; Morgan and Curran 2008). Further research is needed to fully understand the interactions between THC and CBD and the impact of THC:CBD ratios on the beneficial and harmful effects of cannabis.

Marijuana as a Legal Drug

The current state of evidence about the risks and benefits of marijuana is far from optimal, owing to the lack of human clinical trials with a drug that has been illegal for many years. However, on the basis of what we know, cannabis is certainly no more harmful to individuals than many other legally available drugs. As Curran et al. (2016) stated in an important review, "With hindsight, we can clearly see the enormous problems that have been caused to many individuals and to society by tobacco and alcohol. Unlike cannabis, these drugs are legal in most countries, despite the fact that, if asked to decide today which psychoactive drugs should be legal, cannabis (which rarely kills people) might well be judged as being comparatively benign." The current restrictions on cannabis and cannabis products hinder researchers who are attempting to understand their risks and benefits. A carefully regulated market may increase control of the age of initiation of use, provide a more accurate description of dosage, and increase the availability of more balanced cannabis with higher levels of CBD that would reduce the incidence of harm (Curran et al. 2016).

References

Abel, E. L. 1980. *Marihuana: The First Twelve Thousand Years*. Springer.

Abood, M. E., Rizvi, G., Sallapudi, N., and McAllister, S. D. 2001. Activation of the CB1 cannabinoid receptor protects cultured mouse spinal neurons against excitotoxicity. *Neuroscience Letters* 309 (3), 197–201.

Abrahamov, A., Abrahamov, A., and Mechoulam, R. 1995. An efficient new cannabinoid antiemetic in pediatric oncology. *Life Sciences* 56 (23–24), 2097–2102.

Abrams, D. I., Couey, P., Shade, S. B., Kelly, M. E., and Benowitz, N. L. 2011. Cannabinoid-opioid interaction in chronic pain. *Clinical Pharmacology and Therapeutics* 90 (6), 844–851. doi: 10.1038/clpt.2011.188

Abrams, D. I., and Guzman, M. 2015. Cannabis in cancer care. *Clinical Pharmacology and Therapeutics* 97 (6), 575–586. doi: 10.1002/cpt.108

Abrams, D. I., Jay, C. A., Shade, S. B., Vizoso, H., Reda, H., Press, S., et al. 2007. Cannabis in painful HIV-associated sensory neuropathy: A randomized placebo-controlled trial. *Neurology* 68 (7), 515–521. doi: 10.1212/01.wnl .0000253187.66183.9c

Adams, R. 1941. Marihuana. *Harvey Lectures* 37, 168–197.

Adams, R., Harfenist, M., and Loewe, S. 1949. New analogs of tetrahydro-cannabinol. XIX. *Journal of the American Chemical Society* 71, 1624–1628.

Ahmed, A., van der Marck, M. A., van den Elsen, G., and Olde Rikkert, M. 2015. Cannabinoids in late-onset Alzheimer's disease. *Clinical Pharmacology and Therapeutics* 97 (6), 597–606. doi: 10.1002/cpt.117

Ahmed, A., and Wicklund, M. P. 2011. Amyotrophic lateral sclerosis: What role does environment play? *Neurologic Clinics* 29 (3), 689–711. doi: 10.1016/j. ncl.2011.06.001

Ahn, K., Johnson, D. S., Mileni, M., Beidler, D., Long, J. Z., McKinney, M. K., et al. 2009. Discovery and characterization of a highly selective FAAH inhibitor that reduces inflammatory pain. *Chemistry & Biology* 16 (4), 411–420.

Ahn, K., Smith, S. E., Liimatta, M. B., Beidler, D., Sadagopan, N., Dudley, D. T., et al. 2011. Mechanistic and pharmacological characterization of PF-04457845: A highly potent and selective fatty acid amide hydrolase inhibitor that reduces inflammatory and noninflammatory pain. *Journal of Pharmacology and Experimental Therapeutics* 338 (1), 114–124. doi: 10.1124/jpet.111.180257

Alger, B. E. 2012. Endocannabinoids at the synapse a decade after the dies mirabilis (29 March 2001): What we still do not know. *Journal of Physiology* 590 (10), 2203–2212. doi: 10.1113/jphysiol.2011.220855

Alkaitis, M. S., Solorzano, C., Landry, R. P., Piomelli, D., DeLeo, J. A., and Romero-Sandoval, E. A. 2010. Evidence for a role of endocannabinoids, astrocytes and p38 phosphorylation in the resolution of postoperative pain. *PLoS One* 5 (5), e10891. doi: 10.1371/journal.pone.0010891

Allaman, I., Belanger, M., and Magistretti, P. J. 2011. Astrocyte-neuron metabolic relationships: For better and for worse. *Trends in Neurosciences* 34 (2), 76–87. doi: 10.1016/j.tins.2010.12.001

Allen, G. V., Saper, C. B., Hurley, K. M., and Cechetto, D. F. 1991. Organization of visceral and limbic connections in the insular cortex of the rat. *Journal of Comparative Neurology* 311 (1), 1–16.

Allen, J. H., de Moore, G. M., Heddle, R., and Twartz, J. C. 2004. Cannabinoid hyperemesis: Cyclical hyperemesis in association with chronic cannabis abuse. *Gut* 53 (11), 1566–1570. doi: 10.1136/gut.2003.036350

Allsop, D. J., Copeland, J., Lintzeris, N., Dunlop, A. J., Montebello, M., Sadler, C., et al. 2014. Nabiximols as an agonist replacement therapy during cannabis withdrawal: A randomized clinical trial. *JAMA Psychiatry* 71 (3), 281–291. doi: 10.1001/jamapsychiatry.2013.3947

Alpar, A., Di Marzo, V., and Harkany, T. 2016. At the tip of an iceberg: Prenatal marijuana and its possible relation to neuropsychiatric outcome in the offspring. *Biological Psychiatry* 79 (7), e33–e45. doi: 10.1016/j.biopsych.2015.09.009

Alteba, S., Korem, N., & Akirav, I. (2016). Cannabinoids reverse the effects of early stress on neurocognitive performance in adulthood. *Learning & Memory (Cold Spring Harbor, N.Y.)*, 23, 349–358.

Ames, F. R., and Cridland, S. 1986. Anticonvulsant effect of cannabidiol. *South African Medical Journal* 69 (1), 14.

Anderson, S., ed. 2005. *Making Medicines.* Pharmaceutical Press.

Anderson, W. B., Gould, M. J., Torres, R. D., Mitchell, V. A., and Vaughan, C. W. 2014). Actions of the dual FAAH/MAGL inhibitor JZL195 in a murine inflammatory pain model. *Neuropharmacology* 81, 224–230. doi: 10.1016/j.neuropharm.2013.12.018

Andreasson, S., Allebeck, P., Engstrom, A., and Rydberg, U. 1987. Cannabis and schizophrenia: A longitudinal study of Swedish conscripts. *Lancet* 2 (8574), 1483–1486.

Andrews, P. L., and Horn, C. C. 2006. Signals for nausea and emesis: Implications for models of upper gastrointestinal diseases. *Autonomic Neuroscience* 125 (1–2), 100–115.

Andries, A., Frystyk, J., Flyvbjerg, A., and Stoving, R. K. 2014. Dronabinol in severe, enduring anorexia nervosa: A randomized controlled trial. *International Journal of Eating Disorders* 47 (1), 18–23. doi: 10.1002/eat.22173

Arnone, M., Maruani, J., Chaperon, F., Thiebot, M. H., Poncelet, M., Soubrie, P., et al. 1997. Selective inhibition of sucrose and ethanol intake by SR 141716, an antagonist of central cannabinoid (CB1) receptors. *Psychopharmacology* 132 (1), 104–106.

Arseneault, L., Cannon, M., Witton, J., and Murray, R. M. 2004. Causal association between cannabis and psychosis: Examination of the evidence. *British Journal of Psychiatry* 184, 110–117.

Asbridge, M., Hayden, J. A., and Cartwright, J. L. 2012. Acute cannabis consumption and motor vehicle collision risk: Systematic review of observational studies and meta-analysis. *BMJ* 344, e536. doi: 10.1136/bmj.e536

Aso, F., Sanchez-Pla, A., Vegas-Lozano, E., Maldonado, R., and Ferrer, I. 2015. Cannabis-based medicine reduces multiple pathological processes in AbetaPP/PS1 mice. *Journal of Alzheimer's Disease* 43 (3), 977–991. doi: 10.3233/JAD-141014

Bachhuber, M. A., Saloner, B., Cunningham, C. O., and Barry, C. L. 2014. Medical cannabis laws and opioid analgesic overdose mortality in the United States, 1999–2010. *JAMA Internal Medicine* 174 (10), 1668–1673. doi: 10.1001/jamainternmed.2014.4005

Baker, D., Pryce, G., Croxford, J. L., Brown, P., Pertwee, R. G., Huffman, J. W., et al. 2000. Cannabinoids control spasticity and tremor in a multiple sclerosis model. *Nature* 404 (6773), 84–87. doi: 10.1038/35003583

Baker, D., Pryce, G., Croxford, J. L., Brown, P., Pertwee, R. G., Makriyannis, A., et al. 2001. Endocannabinoids control spasticity in a multiple sclerosis model. *FASEB Journal* 15 (2), 300–302. doi: 10.1096/fj.00-0399fje

Baker, D., Pryce, G., Jackson, S. J., Bolton, C., and Giovannoni, G. 2012. The biology that underpins the therapeutic potential of cannabis-based medicines for the control of spasticity in multiple sclerosis. *Multiple Sclerosis and Related Disorders* 1 (2), 64–75. doi: 10.1016/j.msard.2011.11.001

Barkus, E., Morrison, P. D., Vuletic, D., Dickson, J. C., Ell, P. J., Pilowsky, L. S., et al. 2011. Does intravenous Δ^9-tetrahydrocannabinol increase dopamine release? A SPET study. *Journal of Psychopharmacology* 25 (11), 1462–1468. doi: 10.1177/0269881110382465

Bar-Sela, G., Vorobeichik, M., Drawsheh, S., Omer, A., Goldberg, V., and Muller, E. 2013. The medical necessity for medicinal cannabis: Prospective, observational study evaluating the treatment in cancer patients on supportive or palliative care. *Evidence-Based Complementary and Alternative Medicine* 510392. doi: 10.1155/2013/510392

Bass, C. E., and Martin, B. R. 2000. Time course for the induction and maintenance of tolerance to Δ^9-tetrahydrocannabinol in mice. *Drug and Alcohol Dependence* 60 (2), 113–119.

Bauer, M., Chicca, A., Tamborrini, M., Eisen, D., Lerner, R., Lutz, B., et al. 2012. Identification and quantification of a new family of peptide endocannabinoids (pepcans) showing negative allosteric modulation at CB1 receptors. *Journal of Biological Chemistry* 287 (44), 36944–36967. doi: 10.1074/jbc.M112.382481

Beal, B. R., and Wallace, M. S. 2016. An overview of pharmacologic management of chronic pain. *Medical Clinics of North America* 100 (1), 65–79. doi: 10.1016/j.mcna.2015.08.006

Beal, J. E., Olson, R., Lefkowitz, L., Laubenstein, L., Bellman, P., Yangco, B., et al. 1997. Long-term efficacy and safety of dronabinol for acquired immunodeficiency syndrome-associated anorexia. *Journal of Pain and Symptom Management* 14 (1), 7–14. doi: 10.1016/S0885-3924(97)00038-9

Bedard, M., Dubois, S., and Weaver, B. 2007. The impact of cannabis on driving. *Canadian Journal of Public Health* 98 (1), 6–11.

Behonick, G., Shanks, K. G., Firchau, D. J., Mathur, G., Lynch, C. F., Nashelsky, M., et al. 2014. Four postmortem case reports with quantitative detection of the synthetic cannabinoid, 5F-PB-22. *Journal of Analytical Toxicology* 38 (8), 559–562. doi: 10.1093/jat/bku048

Bergamaschi, M. M., Karschner, E. L., Goodwin, R. S., Scheidweiler, K. B., Hirvonen, J., Queiroz, R. H., et al. 2013. Impact of prolonged cannabinoid excretion in chronic daily cannabis smokers' blood on per se drugged driving laws. *Clinical Chemistry* 59 (3), 519–526. doi: 10.1373/clinchem.2012.195503

Bergamaschi, M. M., Queiroz, R. H., Chagas, M. H., de Oliveira, D. C., De Martinis, B. S., Kapczinski, F., et al. 2011. Cannabidiol reduces the anxiety induced by simulated public speaking in treatment-naive social phobia patients. *Neuropsychopharmacology* 36 (6), 1219–1226. doi: 10.1038/npp.2011.6

Bergamaschi, M. M., Queiroz, R. H., Zuardi, A. W., and Crippa, J. A. 2011. Safety and side effects of cannabidiol, a *Cannabis sativa* constituent. *Current Drug Safety* 6 (4), 237–249.

Bergman, J., Delatte, M. S., Paronis, C. A., Vemuri, K., Thakur, G. A., and Makriyannis, A. 2008. Some effects of CB1 antagonists with inverse agonist and neutral biochemical properties. *Physiology & Behavior* 93 (4–5), 666–670. doi: 10.1016/j.physbeh.2007.11.007

Bhaskaran, M. D., and Smith, B. N. 2010. Cannabinoid-mediated inhibition of recurrent excitatory circuitry in the dentate gyrus in a mouse model of temporal lobe epilepsy. *PLoS One* 5 (5), e10683. doi: 10.1371/journal.pone.0010683

Bhattacharyya, S., Atakan, Z., Martin-Santos, R., Crippa, J. A., Kambeitz, J., Prata, D., et al. 2012. Preliminary report of biological basis of sensitivity to the effects of cannabis on psychosis: AKT1 and DAT1 genotype modulates the effects of δ-9-tetrahydrocannabinol on midbrain and striatal function. *Molecular Psychiatry* 17 (12), 1152–1155. doi: 10.1038/mp.2011.187

Bhattacharyya, S., Morrison, P. D., Fusar-Poli, P., Martin-Santos, R., Borgwardt, S., Winton-Brown, T., et al. 2010. Opposite effects of delta-9-tetrahydrocannabinol and cannabidiol on human brain function and psychopathology. *Neuropsychopharmacology* 35 (3), 764–774. doi: 10.1038/npp.2009.184

Bilsland, L. G., Dick, J. R., Pryce, G., Petrosino, S., Di Marzo, V., Baker, D., et al. 2006. Increasing cannabinoid levels by pharmacological and genetic manipulation delay disease progression in SOD1 mice. *FASEB Journal* 20 (7), 1003–1005. doi: 10.1096/fj.05-4743fje

Bisogno, T., Burston, J. J., Rai, R., Allara, M., Saha, B., Mahadevan, A., et al. 2009. Synthesis and pharmacological activity of a potent inhibitor of the biosynthesis of the endocannabinoid 2-arachidonoylglycerol. *ChemMedChem* 4 (6), 946–950. doi: 10.1002/cmdc.200800442

Bisogno, T., Hanus, L., De Petrocellis, L., Tchilibon, S., Ponde, D. E., Brandi, I., et al. 2001. Molecular targets for cannabidiol and its synthetic analogues: Effect on vanilloid VR1 receptors and on the cellular uptake and enzymatic hydrolysis of anandamide. *British Journal of Pharmacology* 134 (4), 845–852. doi: 10.1038/sj.bjp.0704327

Bisogno, T., Mahadevan, A., Coccurello, R., Chang, J. W., Allara, M., Chen, Y., et al. 2013. A novel fluorophosphonate inhibitor of the biosynthesis of the endocannabinoid 2-arachidonoylglycerol with potential anti-obesity effects. *British Journal of Pharmacology* 169 (4), 784–793. doi: 10.1111/bph.12013

Blair, R. E., Deshpande, L. S., and DeLorenzo, R. J. 2015. Cannabinoids: Is there a potential treatment role in epilepsy? *Expert Opinion on Pharmacotherapy* 16 (13), 1911–1914.

Blake, D. R., Robson, P., Ho, M., Jubb, R. W., & McCabe, C. S. (2006). Preliminary assessment of the efficacy, tolerability and safety of a cannabis-based mediciene (Sativex) in the treatment of pain caused by rheumatoid arthritis. *Rheumatology*, 45, 50–52.

Blankman, J. L., Simon, G. M., and Cravatt, B. F. 2007. A comprehensive profile of brain enzymes that hydrolyze the endocannabinoid 2-arachidonoylglycerol. *Chemistry & Biology* 14 (12), 1347–1356. doi: 10.1016/j.chembiol.2007.11.006

Blazquez, C., Chiarlone, A., Sagredo, O., Aguado, T., Pazos, M. R., Resel, E., et al. 2011. Loss of striatal type 1 cannabinoid receptors is a key pathogenic factor in Huntington's disease. *Brain* 134 (1), 119–136. doi: 10.1093/brain/awq278

Bloomfield, M. A., Morgan, C. J., Egerton, A., Kapur, S., Curran, H. V., and Howes, O. D. 2014. Dopaminergic function in cannabis users and its relationship to cannabis-induced psychotic symptoms. *Biological Psychiatry* 75 (6), 470–478. doi: 10.1016/j.biopsych.2013.05.027

Bluett, R. J., Gamble-George, J. C., Hermanson, D. J., Hartley, N. D., Marnett, L. J., and Patel, S. 2014. Central anandamide deficiency predicts stress-induced anxiety: Behavioral reversal through endocannabinoid augmentation. *Translational Psychiatry* 4, e408. doi: 10.1038/tp.2014.53

Bluher, M., Engeli, S., Kloting, N., Berndt, J., Fasshauer, M., Batkai, S., et al. 2006. Dysregulation of the peripheral and adipose tissue endocannabinoid system in human abdominal obesity. *Diabetes* 55 (11), 3053–3060. doi: 10.2337/db06-0812

Boggan, W. O., Steele, R. A., and Freedman, D. X. 1973. Δ⁹-tetrahydrocannabinol effect on audiogenic seizure susceptibility. *Psychopharmacology* 29 (2), 101–106.

Bolognini, D., Rock, E. M., Cluny, N. L., Cascio, M. G., Limebeer, C. L., Duncan, M., et al. 2013. Cannabidiolic acid prevents vomiting in Suncus murinus and nausea-induced behaviour in rats by enhancing 5–HT1A receptor activation. *British Journal of Pharmacology* 168 (6), 1456–1470. doi: 10.1111/bph.12043

Bonnet, U., Specka, M., Stratmann, U., Ochwadt, R., and Scherbaum, N. 2014. Abstinence phenomena of chronic cannabis-addicts prospectively monitored during controlled inpatient detoxification: Cannabis withdrawal syndrome and its correlation with delta-9-tetrahydrocannabinol and -metabolites in serum. *Drug and Alcohol Dependence* 143, 189–197. doi: 10.1016/j.drugalcdep.2014.07.027

Bonn-Miller, M. O., Vujanovic, A. A., and Drescher, K. D. 2011. Cannabis use among military veterans after residential treatment for posttraumatic stress disorder. *Psychology of Addictive Behaviors* 25 (3), 485–491. doi: 10.1037/a0021945

Bossong, M. G., Mehta, M. A., van Berckel, B. N., Howes, O. D., Kahn, R. S., and Stokes, P. R. 2015. Further human evidence for striatal dopamine release induced by administration of Δ⁹-tetrahydrocannabinol (THC): Selectivity to limbic striatum. *Psychopharmacology* 232 (15), 2723–2729. doi: 10.1007/s00213-015-3915-0

Bossong, M. G., van Berckel, B. N., Boellaard, R., Zuurman, L., Schuit, R. C., Windhorst, A. D., et al. 2009. Delta 9-tetrahydrocannabinol induces dopamine release in the human striatum. *Neuropsychopharmacology* 34 (3), 759–766. doi: 10.1038/npp.2008.138

Bouchard, J., Truong, J., Bouchard, K., Dunkelberger, D., Desrayaud, S., Moussaoui, S., et al. 2012. Cannabinoid receptor 2 signaling in peripheral immune cells modulates disease onset and severity in mouse models of Huntington's disease. *Journal of Neuroscience* 32 (50), 18259–18268. doi: 10.1523/JNEUROSCI.4008-12.2012

Bouton, M. E. 2002. Context, ambiguity, and unlearning: Sources of relapse after behavioral extinction. *Biological Psychiatry* 52 (10), 976–986.

Bovasso, G. B. 2001. Cannabis abuse as a risk factor for depressive symptoms. *American Journal of Psychiatry* 158 (12), 2033–2037. doi: 10.1176/appi.ajp.158.12.2033

Braakman, H. M., van Oostenbrugge, R. J., van Kranen-Mastenbroek, V. H., and de Krom, M. C. 2009. Rimonabant induces partial seizures in a patient with a history of generalized epilepsy. *Epilepsia* 50 (9), 2171–2172. doi: 10.1111/j.1528-1167.2009.02203.x

Brecher, E. M. 1972. *Licit and Illicit Drugs*. Little, Brown.

Broyd, S. J., van Hell, H. H., Beale, C., Yucel, M., and Solowij, N. 2016. Acute and chronic effects of cannabinoids on human cognition: A systematic review. *Biological Psychiatry* 79 (7), 557–567. doi: 10.1016/j.biopsych.2015.12.002

Bryden, L., Nicholson, J., Doods, H., & Pekcec, A. (2015). Deficits in spontaneous burrowing behavior in the rat bilateral 66 monosodium iodoacetiate model of osteroarthritis: An objective measure of pain-related behavior and analgesic efficacy. *Osteoarthritis and Cartilage*, 23, 1605–1612.

Burdyga, G., Lal, S., Varro, A., Dimaline, R., Thompson, D. G., and Dockray, G. J. 2004. Expression of cannabinoid CB1 receptors by vagal afferent neurons is inhibited by cholecystokinin. *Journal of Neuroscience* 24 (11), 2708–2715. doi: 10.1523/JNEUROSCI.5404-03.2004

Burston, J. J., Sagar, D. R., Shao, P., Bai, M., King, E., Brailsford, L., et al. 2013. Cannabinoid CB2 receptors regulate central sensitization and pain responses associated with osteoarthritis of the knee joint. *PLoS One* 8 (11), e80440. doi: 10.1371/journal.pone.0080440

Busquets-Garcia, A., Puighermanal, E., Pastor, A., de la Torre, R., Maldonado, R., and Ozaita, A. 2011. Differential role of anandamide and 2-arachidonoylglycerol in memory and anxiety-like responses. *Biological Psychiatry* 70 (5), 479–486. doi: 10.1016/j.biopsych.2011.04.022

Campolongo, P., Roozendaal, B., Trezza, V., Cuomo, V., Astarita, G., Fu, J., et al. 2009. Fat-induced satiety factor oleoylethanolamide enhances memory consolidation. *Proceedings of the National Academy of Sciences* 106 (19), 8027–8031. doi: 10.1073/pnas.0903038106

Campolongo, P., Roozendaal, B., Trezza, V., Hauer, D., Schelling, G., McGaugh, J. L., et al. 2009. Endocannabinoids in the rat basolateral amygdala enhance memory consolidation and enable glucocorticoid modulation of memory. *Proceedings of the National Academy of Sciences* 106 (12), 4888–4893. doi: 10.1073/pnas.0900835106

Campos, A. C., and Guimaraes, F. S. 2008. Involvement of 5HT1A receptors in the anxiolytic-like effects of cannabidiol injected into the dorsolateral periaqueductal gray of rats. *Psychopharmacology* 199 (2), 223–230. doi: 10.1007/s00213-008-1168-x

Campos, A. C., and Guimaraes, F. S. 2009. Evidence for a potential role for TRPV1 receptors in the dorsolateral periaqueductal gray in the attenuation of the anxiolytic effects of cannabinoids. *Progress in Neuro-Psychopharmacology & Biological Psychiatry* 33 (8), 1517–1521. doi: 10.1016/j.pnpbp.2009.08.017

Cannabis-In-Cachexia-Study-Group, Strasser, F., Luftner, D., Possinger, K., Ernst, G., Ruhstaller, T., et al. (2006). Comparison of orally administered cannabis extract and delta-9-tetrahydrocannabinol in treating patients with cancer-related anorexia-cachexia syndrome: A multicenter, phase III, randomized, double-blind, placebo-controlled clinical trial from the Cannabis-In-Cachexia-Study-Group. *Journal of Clinical Oncology*, 24(21), 3394–3400. doi:10.1200/JCO.2005.05.1847.

Cardinal, P., Bellocchio, L., Clark, S., Cannich, A., Klugmann, M., Lutz, B., et al. 2012. Hypothalamic CB1 cannabinoid receptors regulate energy balance in mice. *Endocrinology* 153 (9), 4136–4143. doi: 10.1210/en.2012-1405

Carter, G. T., Weydt, P., Kyashna-Tocha, M., and Abrams, D. I. 2004. Medicinal cannabis: Rational guidelines for dosing. *IDrugs* 7 (5), 464–470.

Caspi, A., Moffitt, T. E., Cannon, M., McClay, J., Murray, R., Harrington, H., et al. 2005. Moderation of the effect of adolescent-onset cannabis use on adult psychosis by a functional polymorphism in the catechol-O-methyltransferase gene: Longitudinal evidence of a gene X environment interaction. *Biological Psychiatry* 57 (10), 1117–1127. doi: 10.1016/j.biopsych.2005.01.026

Castle, D. J. 2013. Cannabis and psychosis: What causes what? *F1000 Medicine Reports* 5, 1. doi: 10.3410/M5-1

Cechetto, D. F., and Saper, C. B. 1987. Evidence for a viscerotopic sensory representation in the cortex and thalamus in the rat. *Journal of Comparative Neurology* 262 (1), 27–45.

Chaperon, F., Soubrie, P., Puech, A. J., and Thiebot, M. H. 1998. Involvement of central cannabinoid (CB1) receptors in the establishment of place conditioning in rats. *Psychopharmacology* 135 (4), 324–332.

Cheah, B. C., Vucic, S., Krishnan, A. V., and Kiernan, M. C. 2010. Riluzole, neuroprotection and amyotrophic lateral sclerosis. *Current Medicinal Chemistry* 17 (18), 1942–1959.

Cheetham, A., Allen, N. B., Whittle, S., Simmons, J. G., Yucel, M., and Lubman, D. I. 2012. Orbitofrontal volumes in early adolescence predict initiation of cannabis use: A 4-year longitudinal and prospective study. *Biological Psychiatry* 71 (8), 684–692. doi: 10.1016/j.biopsych.2011.10.029

Chen, J. J., Ondo, W. G., Dashtipour, K., and Swope, D. M. 2012. Tetrabenazine for the treatment of hyperkinetic movement disorders: A review of the literature. *Clinical Therapeutics* 34 (7), 1487–1504. doi: 10.1016/j.clinthera.2012.06.010

Chen, R., Zhang, J., Fan, N., Teng, Z. Q., Wu, Y., Yang, H., et al. 2013. Delta9-THC-caused synaptic and memory impairments are mediated through COX-2 signaling. *Cell* 155 (5), 1154–1165. doi: 10.1016/j.cell.2013.10.042

Chesher, G. B., Jackson, D. M., and Malor, R. M. 1975. Interaction of delta9-tetrahydrocannabinol and cannabidiol with phenobarbitone in protecting mice from electrically induced convulsions. *Journal of Pharmacy and Pharmacology* 27 (8), 608–609.

Chesher, G. B., Jackson, D. M., and Starmer, G. A. 1974. Interaction of cannabis and general anaesthetic agents in mice. *British Journal of Pharmacology* 50 (4), 593–599.

Chhatwal, J. P., Davis, M., Maguschak, K. A., and Ressler, K. J. 2005. Enhancing cannabinoid neurotransmission augments the extinction of conditioned fear. *Neuropsychopharmacology* 30 (3), 516–524. doi: 10.1038/sj.npp.1300655

Chouker, A., Kaufmann, I., Kreth, S., Hauer, D., Feuerecker, M., Thieme, D., et al. 2010. Motion sickness, stress and the endocannabinoid system. *PLoS One* 5 (5), e10752. doi: 10.1371/journal.pone.0010752

Christensen, R., Kristensen, P. K., Bartels, E. M., Bliddal, H., and Astrup, A. 2007. Efficacy and safety of the weight-loss drug rimonabant: A meta-analysis of randomised trials. *Lancet* 370 (9600), 1706–1713. doi: 10.1016/S0140-6736(07)61721-8

Chrousos, G. P. 2009. Stress and disorders of the stress system. *Nature Reviews. Endocrinology* 5 (7), 374–381. doi: 10.1038/nrendo.2009.106

Clapper, J. R., Moreno-Sanz, G., Russo, R., Guijarro, A., Vacondio, F., Duranti, A., et al. 2010. Anandamide suppresses pain initiation through a peripheral endocannabinoid mechanism. *Nature Neuroscience* 13 (10), 1265–1270. doi: 10.1038/nn.2632

Colasanti, B. K., Lindamood, C., and Craig, C. R. 1982. Effects of marihuana cannabinoids on seizure activity in cobalt-epileptic rats. *Pharmacology, Biochemistry, and Behavior* 16 (4), 573–578.

Comelli, F., Giagnoni, G., Bettoni, I., Colleoni, M., and Costa, B. 2008. Antihyperalgesic effect of a Cannabis sativa extract in a rat model of neuropathic pain: Mechanisms involved. *Phytotherapy Researc*, 22 (8), 1017–1024. doi: 10.1002/ptr.2401

Compston, A., and Coles, A. 2008. Multiple sclerosis. *Lancet* 372 (9648), 1502–1517. doi: 10.1016/S0140-6736(08)61620-7

Consroe, P., Laguna, J., Allender, J., Snider, S., Stern, L., Sandyk, R., et al. 1991. Controlled clinical trial of cannabidiol in Huntington's disease. *Pharmacology, Biochemistry, and Behavior* 40 (3), 701–708.

Consroe, P., Musty, R., Rein, J., Tillery, W., and Pertwee, R. 1997. The perceived effects of smoked cannabis on patients with multiple sclerosis. *European Neurology* 38 (1), 44–48.

Consroe, P. F., Wood, G. C., and Buchsbaum, H. 1975. Anticonvulsant nature of marihuana smoking. *Journal of the American Medical Association* 234 (3), 306–307.

Contreras, M., Ceric, F., and Torrealba, F. 2007. Inactivation of the interoceptive insula disrupts drug craving and malaise induced by lithium. *Science* 318 (5850), 655–658.

Corcoran, L., Roche, M., and Finn, D. P. 2015. The role of the brain's endocannabinoid system in pain and its modulation by stress. *International Review of Neurobiology* 125, 203–255. doi: 10.1016/bs.irn.2015.10.003

Costa, B., and Comelli, F. 2014. Pain. In *Handbook of Cannabis*, ed. R. G. Pertwee. Oxford University Press.

Costa, B., Giagnoni, G., Franke, C., Trovato, A. E., and Colleoni, M. 2004. Vanilloid TRPV1 receptor mediates the antihyperalgesic effect of the nonpsychoactive cannabinoid, cannabidiol, in a rat model of acute inflammation. *British Journal of Pharmacology* 143 (2), 247–250. doi: 10.1038/sj.bjp.0705920

Costall, B., Domeney, A. M., Naylor, R. J., and Tattersall, F. D. 1986. 5-Hydroxytryptamine M-receptor antagonism to prevent cisplatin-induced emesis. *Neuropharmacology* 25 (8), 959–961.

Costas, J., Sanjuan, J., Ramos-Rios, R., Paz, E., Agra, S., Tolosa, A., et al. 2011. Interaction between COMT haplotypes and cannabis in schizophrenia: A case-only study in two samples from Spain. *Schizophrenia Research* 127 (1–3), 22–27. doi: 10.1016/j.schres.2011.01.014

Cox, M. L., Haller, V. L., and Welch, S. P. 2007. The antinociceptive effect of Delta 9-tetrahydrocannabinol in the arthritic rat invovles the CB_2 cannabinoid receptor. *European Journal of Pharmacology* 570 (1–3), 50–56.

Cravatt, B. F., Giang, D. K., Mayfield, S. P., Boger, D. L., Lerner, R. A., and Gilula, N. B. 1996. Molecular characterization of an enzyme that degrades neuromodulatory fatty-acid amides. *Nature* 384 (6604), 83–87.

Crews, F., He, J., and Hodge, C. 2007. Adolescent cortical development: A critical period of vulnerability for addiction. *Pharmacology, Biochemistry, and Behavior* 86 (2), 189–199. doi: 10.1016/j.pbb.2006.12.001

Cridge, B. J., and Rosengren, R. J. 2013. Critical appraisal of the potential use of cannabinoids in cancer management. *Cancer Management and Research* 5, 301–313. doi: 10.2147/CMAR.S36105

Cristino, L., and Di Marzo, V. 2015. Established and emerging conceptos of cannabinoid action on food intake and their potential application to the treatment of anorexia and cachexia. In *Handbook of Cannabis*, ed. R. G. Pertwee. Oxford University Press.

Cross-Mellor, S. K., Ossenkopp, K. P., Piomelli, D., and Parker, L. A. 2007. Effects of the FAAH inhibitor, URB597, and anandamide on lithium-induced taste reactivity responses: A measure of nausea in the rat. *Psychopharmacology* 190 (2), 135–143.

Cunha, J. M., Carlini, E. A., Pereira, A. E., Ramos, O. L., Pimentel, C., Gagliardi, R., et al. 1980. Chronic administration of cannabidiol to healthy volunteers and epileptic patients. *Pharmacology* 21 (3), 175–185.

Cunningham, C. L., Martinez-Cerdeno, V., and Noctor, S. C. 2013. Microglia regulate the number of neural precursor cells in the developing cerebral cortex. *Journal of Neuroscience* 33 (10), 4216–4233. doi: 10.1523/JNEUROSCI.3441-12.2013

Curran, H. V., Brignell, C., Fletcher, S., Middleton, P., and Henry, J. 2002. Cognitive and subjective dose-response effects of acute oral Delta 9-tetrahydrocannabinol (THC) in infrequent cannabis users. *Psychopharmacology* 164 (1), 61–70. doi: 10.1007/s00213-002-1169-0

Curran, H. V., Freeman, T. P., Mokrysz, C., Lewis, D. A., Morgan, C. J., and Parsons, L. H. 2016. Keep off the grass? Cannabis, cognition and addiction. *Nature Reviews Neuroscience* 17 (5), 293–306. doi: 10.1038/nrn.2016.28

Curtis, A., Mitchell, I., Patel, S., Ives, N., and Rickards, H. 2009. A pilot study using nabilone for symptomatic treatment in Huntington's disease. *Movement Disorders* 24 (15), 2254–2259. doi: 10.1002/mds.22809

Curtis, A., and Rickards, H. 2006. Nabilone could treat chorea and irritability in Huntington's disease. *Journal of Neuropsychiatry and Clinical Neurosciences* 18 (4), 553–554. doi: 10.1176/jnp.2006.18.4.553

D'Addario, C., Micioni Di Bonaventura, M. V., Pucci, M., Romano, A., Gaetani, S., Ciccocioppo, R., et al. 2014. Endocannabinoid signaling and food addiction. *Neuroscience and Biobehavioral Reviews* 47, 203–224. doi: 10.1016/j.neubiorev.2014.08.008

Darmani, N. A. 2001a. Delta-9-tetrahydrocannabinol differentially suppresses cisplatin-induced emesis and indices of motor function via cannabinoid CB (1) receptors in the least shrew. *Pharmacology, Biochemistry, and Behavior* 69 (1–2), 239–249.

Darmani, N. A. 2001b. Delta (9)-tetrahydrocannabinol and synthetic cannabinoids prevent emesis produced by the cannabinoid CB (1) receptor antagonist/inverse agonist SR 141716A. *Neuropsychopharmacology* 24 (2), 198–203. doi: 10.1016/S0893-133X(00)00197-4

Das, R. K., Kamboj, S. K., Ramadas, M., Yogan, K., Gupta, V., Redman, E., et al. 2013. Cannabidiol enhances consolidation of explicit fear extinction in humans. *Psychopharmacology* 226 (4), 781–792. doi: 10.1007/s00213-012-2955-y

de Boer-Dennert, M., de Wit, R., Schmitz, P. I., Djontono, J., v Beurden, V., Stoter, G., et al. (1997). Patient perceptions of the side-effects of chemotherapy: The influence of 5HT3 antagonists. *British Journal of Cancer*, 76(8), 1055–1061.

Decoster, J., van Os, J., Kenis, G., Henquet, C., Peuskens, J., De Hert, M., et al. 2011. Age at onset of psychotic disorder: Cannabis, BDNF Val66Met, and sex-specific models of gene-environment interaction. *American Journal of Medical Genetics. Part B, Neuropsychiatric Genetics* 156B (3), 363–369. doi: 10.1002/ajmg.b.31174

Degenhardt, L., Hall, W., and Lynskey, M. 2003. Testing hypotheses about the relationship between cannabis use and psychosis. *Drug and Alcohol Dependence* 71 (1), 37–48.

De Marchi, N., De Petrocellis, L., Orlando, P., Daniele, F., Fezza, F., and Di Marzo, V. 2003. Endocannabinoid signalling in the blood of patients with schizophrenia. *Lipids in Health and Disease* 2, 5. doi: 10.1186/1476-511X-2-5

Deng, L., Cornett, B. L., Mackie, K., and Hohmann, A. G. 2015. CB1 knockout mice unveil sustained CB2-mediated antiallodynic effects of the mixed CB1/CB2 agonist CP55,940 in a mouse model of paclitaxel-induced neuropathic pain. *Molecular Pharmacology* 88 (1), 64–74. doi: 10.1124/mol.115.098483

Denson, T. F., and Earleywine, M. 2006. Decreased depression in marijuana users. *Addictive Behaviors* 31 (4), 738–742. doi: 10.1016/j.addbeh.2005.05.052

Derocq, J. M., Bouaboula, M., Marchand, J., Rinaldi-Carmona, M., Segui, M., and Casellas, P. 1998. The endogenous cannabinoid anandamide is a lipid messenger activating cell growth via a cannabinoid receptor-independent pathway in hematopoietic cell lines. *FEBS Letters* 425 (3), 419–425.

Despres, J. P., Golay, A., and Sjöström, L. 2005. Effects of rimonabant on metabolic risk factors in overweight patients with dyslipidemia. *New England Journal of Medicine* 353 (20), 2121–2134. doi: 10.1056/NEJMoa044537

Desroches, J., Guindon, J., Lambert, C., and Beaulieu, P. 2008. Modulation of the anti-nociceptive effects of 2-arachidonoyl glycerol by peripherally administered FAAH and MGL inhibitors in a neuropathic pain model. *British Journal of Pharmacology* 155 (6), 913–924. doi: 10.1038/bjp.2008.322

Devane, W. A., Dysarz, F. A., Johnson, M. R., Melvin, L. S., and Howlett, A. C. 1988. Determination and characterization of a cannabinoid receptor in rat brain. *Molecular Pharmacology* 34 (5), 605–613.

Devane, W. A., Hanus, L., Breuer, A., Pertwee, R. G., Stevenson, L. A., Griffin, G., et al. 1992. Isolation and structure of a brain constituent that binds to the cannabinoid receptor. *Science* 258 (5090), 1946–1949.

Deveaux, V., Cadoudal, T., Ichigotani, Y., Teixeira-Clerc, F., Louvet, A., Manin, S., et al. 2009. Cannabinoid CB2 receptor potentiates obesity-associated inflammation, insulin resistance and hepatic steatosis. *PLoS One* 4 (6), e5844. doi: 10.1371/journal.pone.0005844

Devinsky, O., Cilio, M. R., Cross, H., Fernandez-Ruiz, J., French, J., Hill, C., et al. 2014. Cannabidiol: Pharmacology and potential therapeutic role in epilepsy and other neuropsychiatric disorders. *Epilepsia* 55 (6), 791–802. doi: 10.1111/epi.12631

Devinsky, O., Marsh, E., Friedman, D., Thiele, E., Laux, L., Sullivan, J., et al. 2016. Cannabidiol in patients with treatment-resistant epilepsy: An open-label interventional trial. *Lancet Neurology* 15 (3), 270–278. doi: 10.1016/S1474-4422(15)00379-8

Di Forti, M., Iyegbe, C., Sallis, H., Kolliakou, A., Falcone, M. A., Paparelli, A., et al. 2012. Confirmation that the AKT1 (rs2494732) genotype influences the risk of psychosis in cannabis users. *Biological Psychiatry* 72 (10), 811–816. doi: 10.1016/j.biopsych.2012.06.020

Di Forti, M., Morgan, C., Dazzan, P., Pariante, C., Mondelli, V., Marques, T. R., et al. 2009. High-potency cannabis and the risk of psychosis. *British Journal of Psychiatry* 195 (6), 488–491. doi: 10.1192/bjp.bp.109.064220

Di Forti, M., Morrison, P. D., Butt, A., and Murray, R. M. 2007. Cannabis use and psychiatric and cogitive disorders: The chicken or the egg? *Current Opinion in Psychiatry* 20 (3), 228–234. doi: 10.1097/YCO.0b013e3280fa838e

Di Marzo, V., Blumberg, P. M., and Szallasi, A. 2002. Endovanilloid signaling in pain. *Current Opinion in Neurobiology* 12 (4), 372–379.

Di Marzo, V., Goparaju, S. K., Wang, L., Liu, J., Batkai, S., Jarai, Z., et al. 2001. Leptin-regulated endocannabinoids are involved in maintaining food intake. *Nature* 410 (6830), 822–825. doi: 10.1038/35071088

Dincheva, I., Drysdale, A. T., Hartley, C. A., Johnson, D. C., Jing, D., King, E. C., et al. 2015. FAAH genetic variation enhances fronto-amygdala function in mouse and human. *Nature Communications* 6, 6395. doi: 10.1038/ncomms7395

Dinh, T. P., Kathuria, S., and Piomelli, D. 2004. RNA interference suggests a primary role for monoacylglycerol lipase in the degradation of the endocannabinoid 2-arachidonoylglycerol. *Molecular Pharmacology* 66 (5), 1260–1264.

Dregan, A., and Gulliford, M. C. 2012. Is illicit drug use harmful to cognitive functioning in the midadult years? A cohort-based investigation. *American Journal of Epidemiology* 175 (3), 218–227. doi: 10.1093/aje/kwr315

D'Souza, D. C. 2007. Cannabinoids and psychosis. *International Review of Neurobiology* 78, 289–326. doi: 10.1016/S0074-7742(06)78010-2

D'Souza, D. C., Abi-Saab, W. M., Madonick, S., Forselius-Bielen, K., Doersch, A., Braley, G., et al. 2005. Delta-9-tetrahydrocannabinol effects in schizophrenia: Implications for cognition, psychosis, and addiction. *Biological Psychiatry* 57 (6), 594–608. doi: 10.1016/j.biopsych.2004.12.006

D'Souza, D. C., Cortes-Briones, J. A., Ranganathan, M., Thurnauer, H., Creatura, G., Surti, T., et al. 2016. Rapid changes in CB1 receptor availability in cannabis dependent males after abstinence from cannabis. *Biol Psychiatry Cogn Neurosci Neuroimaging* 1 (1), 60–67. doi: 10.1016/j.bpsc.2015.09.008

D'Souza, D. C., Perry, E., MacDougall, L., Ammerman, Y., Cooper, T., Wu, Y. T., et al. 2004. The psychotomimetic effects of intravenous delta-9-tetrahydrocannabinol in healthy individuals: Implications for psychosis. *Neuropsychopharmacology* 29 (8), 1558–1572. doi: 10.1038/sj.npp.1300496

Duran, M., Perez, E., Abanades, S., Vidal, X., Saura, C., Majem, M., et al. 2010. Preliminary efficacy and safety of an oromucosal standardized cannabis extract in chemotherapy-induced nausea and vomiting. *British Journal of Clinical Pharmacology* 70 (5), 656–663. doi: 10.1111/j.1365-2125.2010.03743.x

Dyson, A., Peacock, M., Chen, A., et al. 2005. Antihyperalgesic properties of the cannabinoid CT-3 in chronic neuropathic and inflammatory pain states in the rat. *Pain* 116 (1–2), 129–137.

Earleywine, M., and Barnwell, S. S. 2007. Decreased respiratory symptoms in cannabis users who vaporize. *Harm Reduction Journal* 4, 11. doi: 10.1186/1477-7517-4-11

Ellis, R. J., Toperoff, W., Vaida, F., van den Brande, G., Gonzales, J., Gouaux, B., et al. 2009. Smoked medicinal cannabis for neuropathic pain in HIV: A randomized, crossover clinical trial. *Neuropsychopharmacology* 34 (3), 672–680. doi: 10.1038/npp.2008.120

El-Remessy, A. B., Khalil, I. E., Matragoon, S., Abou-Mohamed, G., Tsai, N. J., Roon, P., et al. 2003. Neuroprotective effect of (-)Delta9-tetrahydrocannabinol and cannabidiol in N-methyl-D-aspartate-induced retinal neurotoxicity: Involvement of peroxynitrite. *American Journal of Pathology* 163 (5), 1997–2008.

ElSohly, M. A., Mehmedic, Z., Foster, S., Gon, C., Chandra, S., and Church, J. C. 2016. Changes in cannabis potency over the last 2 decades (1995–2014): Analysis of current data in the United States. *Biological Psychiatry* 79 (7), 613–619. doi: 10.1016/j.biopsych.2016.01.004

ElSohly, M., and Gul, W. 2014. Constituents of *Cannabis sativa*. In *Handbook of Cannabis*, ed. R. G. Pertwee. Oxford University Press.

Farrimond, J. A., Whalley, B. J., and Williams, C. M. 2012. Cannabinol and cannabidiol exert opposing effects on rat feeding patterns. *Psychopharmacology* 223 (1), 117–129. doi: 10.1007/s00213-012-2697-x

Fegley, D., Gaetani, S., Duranti, A., Tontini, A., Mor, M., Tarzia, G., et al. 2005. Characterization of the fatty acid amide hydrolase inhibitor cyclohexyl carbamic acid 3′-carbamoyl-biphenyl-3-yl ester (URB597): Effects on anandamide and oleoylethanolamide deactivation. *Journal of Pharmacology and Experimental Therapeutics* 313 (1), 352–358. doi: 10.1124/jpet.104.078980

Feinberg, I., Jones, R., Walker, J., Cavness, C., and Floyd, T. 1976. Effects of marijuana extract and tetrahydrocannabinol on electroencephalographic sleep patterns. *Clinical Pharmacology and Therapeutics* 19 (6), 782–794.

Feinberg, I., Jones, R., Walker, J. M., Cavness, C., and March, J. 1975. Effects of high dosage delta-9-tetrahydrocannabinol on sleep patterns in man. *Clinical Pharmacology and Therapeutics* 17 (4), 458–466.

Fergusson, D. M., Horwood, L. J., and Ridder, E. M. 2005. Tests of causal linkages between cannabis use and psychotic symptoms. *Addiction* 100 (3), 354–366. doi: 10.1111/j.1360-0443.2005.01001.x

Fernandez-Ruiz, J. 2009. The endocannabinoid system as a target for the treatment of motor dysfunction. *British Journal of Pharmacology* 156 (7), 1029–1040. doi: 10.1111/j.1476-5381.2008.00088.x

Fernandez-Ruiz, J., deLago, E., Gomez-Ruiz, M., Garcia, C., Sagredo, O., and Garcia-Arencibia, M. 2014. Neurodegenerative disorders other than multiple sclerosis. In *Handbook of Cannabis*, ed. R. G. Pertwee. Oxford University Press.

Fernandez-Ruiz, J., Garcia, C., Sagredo, O., Gomez-Ruiz, M., and de Lago, E. 2010. The endocannabinoid system as a target for the treatment of neuronal damage. *Expert Opinion on Therapeutic Targets* 14 (4), 387–404. doi: 10.1517/14728221003709792

Fernandez-Ruiz, J., & Gonzales, S. (2005). Cannabinoid control of motor function at the basal ganglia. *Handbook of Experimental Pharmacology*, 168, 479–507.

Fernandez-Ruiz, J., Romero, J., Velasco, G., Tolon, R. M., Ramos, J. A., and Guzman, M. 2007. Cannabinoid CB2 receptor: A new target for controlling neural cell survival? *Trends in Pharmacological Sciences* 28 (1), 39–45. doi: 10.1016/j.tips.2006.11.001

Fernandez-Ruiz, J., Sagredo, O., Pazos, M. R., Garcia, C., Pertwee, R., Mechoulam, R., et al. 2013. Cannabidiol for neurodegenerative disorders: Important new clinical applications for this phytocannabinoid? *British Journal of Clinical Pharmacology* 75 (2), 323–333. doi: 10.1111/j.1365-2125.2012.04341.x

Filbey, F. M., Aslan, S., Calhoun, V. D., Spence, J. S., Damaraju, E., Caprihan, A., et al. 2014. Long-term effects of marijuana use on the brain. *Proceedings of the National Academy of Sciences* 111 (47), 16913–16918. doi: 10.1073/pnas.1415297111

Filbey, F. M., Schacht, J. P., Myers, U. S., Chavez, R. S., and Hutchison, K. E. 2009. Marijuana craving in the brain. *Proceedings of the National Academy of Sciences* 106 (31), 13016–13021. doi: 10.1073/pnas.0903863106

Filloux, F. M. 2015. Cannabinoids for pediatric epilepsy? Up in smoke or real science? *Translational Pediatrics* 4 (4), 271–282. doi: 10.3978/j.issn.2224-4336.2015.10.03

Fine, P. G. 2012. Treatment guidelines for the pharmacological management of pain in older persons. *Pain Medicine* 13 (supplement 2), S57–S66. doi: 10.1111/j.1526-4637.2011.01307.x

Fisher, R. S. 2012. Therapeutic devices for epilepsy. *Annals of Neurology* 71 (2), 157–168. doi: 10.1002/ana.22621

Foltin, R. W., Brady, J. V., and Fischman, M. W. 1986. Behavioral analysis of marijuana effects on food intake in humans. *Pharmacology, Biochemistry, and Behavior* 25 (3), 577–582.

Foltin, R. W., Fischman, M. W., and Byrne, M. F. 1988. Effects of smoked marijuana on food intake and body weight of humans living in a residential laboratory. *Appetite* 11 (1), 1–14.

Forget, B., Coen, K. M., and Le Foll, B. 2009. Inhibition of fatty acid amide hydrolase reduces reinstatement of nicotine seeking but not break point for nicotine self-administration: Comparison with CB (1) receptor blockade. *Psychopharmacology* 205 (4), 613–624. doi: 10.1007/s00213-009-1569-5

Forget, B., Guranda, M., Gamaleddin, I., Goldberg, S. R., and Le Foll, B. 2016. Attenuation of cue-induced reinstatement of nicotine seeking by URB597 through cannabinoid CB receptor in rats. *Psychopharmacology*. doi: 10.1007/s00213-016-4232-y

Formukong, E. A., Evans, A. T., and Evans, F. J. 1988. Analgesic and antiinflammatory activity of constituents of *Cannabis sativa* L. *Inflammation* 12 (4), 361–371.

Fowler, C. J. 2015. The potential of inhibitors of endocannabinoid metabolism for drug development: A critical review. In R. G. Pertwee (Ed.), *Endocannabinoids*. Springer.

Fraser, G. A. 2009. The use of a synthetic cannabinoid in the management of treatment-resistant nightmares in posttraumatic stress disorder (PTSD). *CNS Neuroscience & Therapeutics* 15 (1), 84–88. doi: 10.1111/j.1755-5949.2008.00071.x

Freeman, T. P., Morgan, C. J., Hindocha, C., Schafer, G., Das, R. K., and Curran, H. V. 2014. Just say 'know': How do cannabinoid concentrations influence users' estimates of cannabis potency and the amount they roll in joints? *Addiction* 109 (10), 1686–1694. doi: 10.1111/add.12634

Fride, E., Ginzburg, Y., Breuer, A., Bisogno, T., Di Marzo, V., and Mechoulam, R. 2001. Critical role of the endogenous cannabinoid system in mouse pup suckling and growth. *European Journal of Pharmacology* 419 (2–3), 207–214.

Fried, P. A., Watkinson, B., and Gray, R. 2005. Neurocognitive consequences of marihuana—a comparison with pre-drug performance. *Neurotoxicology and Teratology* 27 (2), 231–239. doi: 10.1016/j.ntt.2004.11.003

Frieling, H., Albrecht, H., Jedtberg, S., Gozner, A., Lenz, B., Wilhelm, J., et al. 2009. Elevated cannabinoid 1 receptor mRNA is linked to eating disorder related behavior and attitudes in females with eating disorders. *Psychoneuroendocrinology* 34 (4), 620–624. doi: 10.1016/j.psyneuen.2008.10.014

Frisher, M., Crome, I., Martino, O., and Croft, P. 2009. Assessing the impact of cannabis use on trends in diagnosed schizophrenia in the United Kingdom from 1996 to 2005. *Schizophrenia Research* 113 (2–3), 123–128. doi: 10.1016/j. schres.2009.05.031

Fu, J., Bottegoni, G., Sasso, O., Bertorelli, R., Rocchia, W., Masetti, M., et al. 2012. A catalytically silent FAAH-1 variant drives anandamide transport in neurons. *Nature Neuroscience* 15 (1), 64–69.

Fusar-Poli, P., Crippa, J. A., Bhattacharyya, S., Borgwardt, S. J., Allen, P., Martin-Santos, R., et al. 2009. Distinct effects of {delta}9-tetrahydrocannabinol and cannabidiol on neural activation during emotional processing. *Archives of General Psychiatry* 66 (1), 95–105. doi: 10.1001/archgenpsychiatry.2008.519

Fuss, J., Steinle, J., Bindila, L., Auer, M. K., Kirchherr, H., Lutz, B., et al. (2015). A runner's high depends on cannabinoid receptors in mice. *Proceedings of the National Academy of Sciences of the United States of America*, 112(42), 13105–13108. doi:10.1073/pnas.151996112.

Gage, S. H., Hickman, M., and Zammit, S. 2016. Association between cannabis and psychosis: Epidemiologic evidence. *Biological Psychiatry* 79 (7), 549–556. doi: 10.1016/j.biopsych.2015.08.001

Gage, S. H., Munafo, M. R., MacLeod, J., Hickman, M., and Smith, G. D. 2015. Cannabis and psychosis. *Lancet. Psychiatry* 2 (5), 380. doi: 10.1016/S2215-0366(15)00108-X

Gage, S. H., Zammit, S., and Hickman, M. 2013. Stronger evidence is needed before accepting that cannabis plays an important role in the aetiology of schizophrenia in the population. *F1000 Medicine Reports* 5, 2. doi: 10.3410/ M5-2

Gallate, J. E., and McGregor, I. S. 1999. The motivation for beer in rats: Effects of ritanserin, naloxone and SR 141716. *Psychopharmacology* 142 (3), 302–308.

Gallate, J. E., Saharov, T., Mallet, P. E., and McGregor, I. S. 1999. Increased motivation for beer in rats following administration of a cannabinoid CB1 receptor agonist. *European Journal of Pharmacology* 370 (3), 233–240.

Gaoni, Y., and Mechoulam, R. 1964a. Isolation, structure, and partial synthesis of an active constituent of hashish. *Journal of the American Chemical Society* 86, 1646–1647.

Gaoni, Y., & Mechoulam, R. 1964b. The structure and synthesis of cannabigerol, a new hashish constituent. *Proceedings of the Chemical Society*, 82.

Garcia, C., Palomo-Garo, C., Garcia-Arencibia, M., Ramos, J., Pertwee, R., and Fernandez-Ruiz, J. 2011. Symptom-relieving and neuroprotective effects of the phytocannabinoid Delta (9)-THCV in animal models of Parkinson's disease. *British Journal of Pharmacology* 163 (7), 1495–1506. doi: 10.1111/j.1476 -5381.2011.01278.x

Garcia-Arencibia, M., Garcia, C., and Fernandez-Ruiz, J. 2009. Cannabinoids and Parkinson's disease. *CNS & Neurological Disorders—Drug Targets* 8 (6), 432–439.

Gardner, E. L. 2005. Endocannabinoid signaling system and brain reward: Emphasis on dopamine. *Pharmacology, Biochemistry, and Behavior* 81 (2), 263–284. doi: 10.1016/j.pbb.2005.01.032

Gates, P. J., Albertella, L., and Copeland, J. 2014. The effects of cannabinoid administration on sleep: A systematic review of human studies. *Sleep Medicine Reviews* 18 (6), 477–487. doi: 10.1016/j.smrv.2014.02.005

Gerard, N., Pieters, G., Goffin, K., Bormans, G., and Van Laere, K. 2011. Brain type 1 cannabinoid receptor availability in patients with anorexia and bulimia nervosa. *Biological Psychiatry* 70 (8), 777–784. doi: 10.1016/j.biopsych.2011 .05.010

Gerostamoulos, D., Drummer, O. H., and Woodford, N. W. 2015. Deaths linked to synthetic cannabinoids. *Forensic Science, Medicine, and Pathology* 11 (3), 478. doi: 10.1007/s12024-015-9669-5

Gertsch, J., Leonti, M., Raduner, S., Racz, I., Chen, J. Z., Xie, X. Q., et al. 2008. Beta-caryophyllene is a dietary cannabinoid. *Proceedings of the National Academy of Sciences* 105 (26), 9099–9104. doi: 10.1073/pnas.0803601105

Ghosh, P., and Bhattacharya, S. K. 1978. Anticonvulsant action of cannabis in the rat: Role of brain monoamines. *Psychopharmacology* 59 (3), 293–297.

Ghosh, S., Kinsey, S. G., Liu, Q. S., Hruba, L., McMahon, L. R., Grim, T. W., et al. 2015. Full Fatty Acid Amide Hydrolase Inhibition Combined with Partial Monoacylglycerol Lipase Inhibition: Augmented and Sustained Antinociceptive Effects with Reduced Cannabimimetic Side Effects in Mice. *Journal of Pharmacology and Experimental Therapeutics* 354 (2), 111–120. doi: 10.1124/ jpet.115.222851

Gieringer, D., St Laurent, J., and Goodrich, S. 2004. Cannabis vaporizer combines efficent delivery of THC with effective suppression of pryolytic compounds. *Journal of Cannabis Therapeutics* 4, 7–27.

Gill, E. W. 1971. Propyl homologue of tetrahydrocannabinol: Its isolation from cannabis, properties, and synthesis. *Journal of the Chemical Society, Section C: Organic*, 579–582.

Gilman, J. M., Kuster, J. K., Lee, S., Lee, M. J., Kim, B. W., Makris, N., et al. 2014. Cannabis use is quantitatively associated with nucleus accumbens and amygdala abnormalities in young adult recreational users. *Journal of Neuroscience* 34 (16), 5529–5538. doi: 10.1523/JNEUROSCI.4745-13.2014

Giuffrida, A., Leweke, F. M., Gerth, C. W., Schreiber, D., Koethe, D., Faulhaber, J., et al. 2004. Cerebrospinal anandamide levels are elevated in acute schizophrenia and are inversely correlated with psychotic symptoms. *Neuropsychopharmacology* 29 (11), 2108–2114. doi: 10.1038/sj.npp.1300558

Gomes, F. V., Alves, F. H., Guimaraes, F. S., Correa, F. M., Resstel, L. B., and Crestani, C. C. 2013. Cannabidiol administration into the bed nucleus of the stria terminalis alters cardiovascular responses induced by acute restraint stress through 5-HT (1)A receptor. *European Neuropsychopharmacology* 23 (9), 1096–1104. doi: 10.1016/j.euroneuro.2012.09.007

Gomez, R., Navarro, M., Ferrer, B., Trigo, J. M., Bilbao, A., Del Arco, I., et al. 2002. A peripheral mechanism for CB1 cannabinoid receptor-dependent modulation of feeding. *Journal of Neuroscience* 22 (21), 9612–9617.

Gonzalez, S., Cebeira, M., and Fernandez-Ruiz, J. 2005. Cannabinoid tolerance and dependence: A review of studies in laboratory animals. *Pharmacology, Biochemistry, and Behavior* 81 (2), 300–318. doi: 10.1016/j.pbb.2005.01.028

Goonawardena, A. V., Robinson, L., Hampson, R. E., and Riedel, G. 2010. Cannabinoid and cholinergic systems interact during performance of a short-term memory task in the rat. *Learning & Memory* 17 (10), 502–511. doi: 10.1101/lm.1893710

Gorzalka, B. B., Hill, M. N., and Chang, S. C. 2010. Male-female differences in the effects of cannabinoids on sexual behavior and gonadal hormone function. *Hormones and Behavior* 58 (1), 91–99. doi: 10.1016/j.yhbeh.2009.08.009

Grant, I., Gonzalez, R., Carey, C. L., Natarajan, L., and Wolfson, T. 2003. Non-acute (residual) neurocognitive effects of cannabis use: A meta-analytic study. *Journal of the International Neuropsychological Society* 9 (5), 679–689. doi: 10.1017/S1355617703950016

Gregg, L. C., Jung, K. M., Spradley, J. M., Nyilas, R., Suplita, R. L., II, Zimmer, A., et al. 2012. Activation of type 5 metabotropic glutamate receptors and diacylglycerol lipase-alpha initiates 2-arachidonoylglycerol formation and endocannabinoid-mediated analgesia. *Journal of Neuroscience* 32 (28), 9457–9468. doi: 10.1523/JNEUROSCI.0013-12.2012

Grill, H. J., and Norgren, R. 1978. The taste reactivity test. I. Mimetic responses to gustatory stimuli in neurologically normal rats. *Brain Research* 143 (2), 263–279.

Grinspoon, L., and Bakalar, J. B. 1997. *Marihuana, the Forbidden Medicine.* Yale University Press.

Gross, D. W., Hamm, J., Ashworth, N. L., and Quigley, D. 2004. Marijuana use and epilepsy: Prevalence in patients of a tertiary care epilepsy center. *Neurology* 62 (11), 2095–2097.

Gross, H., Ebert, M. H., Faden, V. B., Goldberg, S. C., Kaye, W. H., Caine, E. D., et al. 1983. A double-blind trial of delta 9-tetrahydrocannabinol in primary anorexia nervosa. *Journal of Clinical Psychopharmacology* 3 (3), 165–171.

Gruber, A. J., Pope, H. G., Jr., and Brown, M. E. 1996. Do patients use marijuana as an antidepressant? *Depression* 4 (2), 77–80. doi: 10.1002/(SICI)1522-71621996)4:2<77:AID-DEPR7>3.0.CO;2-C

Grunfeld, Y., and Edery, H. 1969. Psychopharmacological activity of the active constituents of hashish and some related cannabinoids. *Psychopharmacology* 14 (3), 200–210.

Gubellini, P., Picconi, B., Bari, M., Battista, N., Calabresi, P., Centonze, D., et al. 2002. Experimental parkinsonism alters endocannabinoid degradation: Implications for striatal glutamatergic transmission. *Journal of Neuroscience* 22 (16), 6900–6907. doi: 20026732

Guerry, J. D., and Hastings, P. D. 2011. In search of HPA axis dysregulation in child and adolescent depression. *Clinical Child and Family Psychology Review* 14 (2), 135–160. doi: 10.1007/s10567-011-0084-5

Guimaraes, F. S., Chiaretti, T. M., Graeff, F. G., and Zuardi, A. W. 1990. Antianxiety effect of cannabidiol in the elevated plus-maze. *Psychopharmacology* 100 (4), 558–559.

Guindon, J., Guijarro, A., Piomelli, D., and Hohmann, A. G. 2011. Peripheral antinociceptive effects of inhibitors of monoacylglycerol lipase in a rat model of inflammatory pain. *British Journal of Pharmacology* 163 (7), 1464–1478. doi: 10.1111/j.1476-5381.2010.01192.x

Guindon, J., and Hohmann, A. G. 2009. The endocannabinoid system and pain. *CNS & Neurological Disorders—Drug Targets* 8 (6), 403–421.

Guindon, J., Lai, Y., Takacs, S. M., Bradshaw, H. B., and Hohmann, A. G. 2013. Alterations in endocannabinoid tone following chemotherapy-induced peripheral neuropathy: Effects of endocannabinoid deactivation inhibitors targeting fatty-acid amide hydrolase and monoacylglycerol lipase in comparison to reference analgesics following cisplatin treatment. *Pharmacological Research* 67 (1), 94–109. doi: 10.1016/j.phrs.2012.10.013

Haller, J., Barna, I., Barsvari, B., Gyimesi Pelczer, K., Yasar, S., Panlilio, L. V., et al. 2009. Interactions between environmental aversiveness and the anxiolytic effects of enhanced cannabinoid signaling by FAAH inhibition in rats. *Psychopharmacology* 204 (4), 607–616. doi: 10.1007/s00213-009-1494-7

Hampson, A. J., Grimaldi, M., Lolic, M., Wink, D., Rosenthal, R., and Axelrod, J. 2000. Neuroprotective antioxidants from marijuana. *Annals of the New York Academy of Sciences* 899, 274–282.

Han, J., Kesner, P., Metna-Laurent, M., Duan, T., Xu, L., Georges, F., et al. 2012. Acute cannabinoids impair working memory through astroglial CB1 receptor modulation of hippocampal LTD. *Cell* 148 (5), 1039–1050. doi: 10.1016/j. cell.2012.01.037

Haney, M., Cooper, Z. D., Bedi, G., Vosburg, S. K., Comer, S. D., and Foltin, R. W. 2013. Nabilone decreases marijuana withdrawal and a laboratory measure of marijuana relapse. *Neuropsychopharmacology* 38 (8), 1557–1565. doi: 10.1038/npp.2013.54

Haney, M., Hart, C. L., Vosburg, S. K., Nasser, J., Bennett, A., Zubaran, C., et al. 2004. Marijuana withdrawal in humans: Effects of oral THC or divalproex. *Neuropsychopharmacology* 29 (1), 158–170. doi: 10.1038/sj.npp.1300310

Hansen, H. H., Ikonomidou, C., Bittigau, P., Hansen, S. H., and Hansen, H. S. 2001. Accumulation of the anandamide precursor and other N-acylethanolamine phospholipids in infant rat models of in vivo necrotic and apoptotic neuronal death. *Journal of Neurochemistry* 76 (1), 39–46.

Hansen, H. H., Schmid, P. C., Bittigau, P., Lastres-Becker, I., Berrendero, F., Manzanares, J., et al. 2001. Anandamide, but not 2-arachidonoylglycerol, accumulates during in vivo neurodegeneration. *Journal of Neurochemistry* 78 (6), 1415–1427.

Hanus, L., Breuer, A., Tchilibon, S., Shiloah, S., Goldenberg, D., Horowitz, M., et al. 1999. HU-308: A specific agonist for CB (2), a peripheral cannabinoid receptor. *Proceedings of the National Academy of Sciences* 96 (25), 14228–14233.

Harder, V. S., Morral, A. R., and Arkes, J. 2006. Marijuana use and depression among adults: Testing for causal associations. *Addiction* 101 (10), 1463–1472. doi: 10.1111/j.1360-0443.2006.01545.x

Harrold, J. A., Elliott, J. C., King, P. J., Widdowson, P. S., and Williams, G. 2002. Down-regulation of cannabinoid-1 (CB-1) receptors in specific extrahypothalamic regions of rats with dietary obesity: A role for endogenous cannabinoids in driving appetite for palatable food? *Brain Research* 952 (2), 232–238.

Hartman, R. L., and Huestis, M. A. 2013. Cannabis effects on driving skills. *Clinical Chemistry* 59 (3), 478–492. doi: 10.1373/clinchem.2012.194381

Hayatbakhsh, R., Williams, G. M., Bor, W., and Najman, J. M. 2013. Early childhood predictors of age of initiation to use of cannabis: A birth prospective study. *Drug and Alcohol Review* 32 (3), 232–240. doi: 10.1111/j.1465-3362 .2012.00520.x

Hazekamp, A., Ware, M. A., Muller-Vahl, K. R., Abrams, D., and Grotenhermen, F. 2013. The medicinal use of cannabis and cannabinoids—an international cross-sectional survey on administration forms. *Journal of Psychoactive Drugs* 45 (3), 199–210. doi: 10.1080/02791072.2013.805976

Heifets, B. D., and Castillo, P. E. 2009. Endocannabinoid signaling and long-term synaptic plasticity. *Annual Review of Physiology* 71, 283–306. doi: 10.1146/annurev.physiol.010908.163149

Henquet, C., Di Forti, M., Morrison, P., Kuepper, R., and Murray, R. M. 2008. Gene-environment interplay between cannabis and psychosis. *Schizophrenia Bulletin* 34 (6), 1111–1121. doi: 10.1093/schbul/sbn108

Henquet, C., Rosa, A., Delespaul, P., Papiol, S., Fananas, L., van Os, J., et al. 2009. COMT ValMet moderation of cannabis-induced psychosis: A momentary assessment study of 'switching on' hallucinations in the flow of daily life. *Acta Psychiatrica Scandinavica* 119 (2), 156–160. doi: 10.1111/j.1600-0447.2008 .01265.x

Henquet, C., Rosa, A., Krabbendam, L., Papiol, S., Fananas, L., Drukker, M., et al. 2006. An experimental study of catechol-o-methyltransferase Val158Met moderation of delta-9-tetrahydrocannabinol-induced effects on psychosis and cognition. *Neuropsychopharmacology* 31 (12), 2748–2757. doi: 10.1038/sj.npp.1301197

Henquet, C., and van Os, J. 2008. The coherence of the evidence linking cannabis with psychosis. *Psychological Medicine* 38 (3), 461–462, author reply 462–464. doi: 10.1017/S0033291707002279

Henze, T., Flachenecker, P., and Zettl, U. K. 2013. Bedeutung und Behandlung der Spastik bei Multipler Sklerose: Ergebnisse der MOVE-1-Studie. *Der Nervenarzt* 84 (2), 214–222. doi: 10.1007/s00115-012-3724-1

Herkenham, M., Lynn, A. B., Little, M. D., Johnson, M. R., Melvin, L. S., de Costa, B. R., et al. 1990. Cannabinoid receptor localization in brain. *Proceedings of the National Academy of Sciences* 87 (5), 1932–1936.

Hermann, H., Marsicano, G., and Lutz, B. 2002. Coexpression of the cannabinoid receptor type 1 with dopamine and serotonin receptors in distinct neuronal subpopulations of the adult mouse forebrain. *Neuroscience* 109 (3), 451–460.

Hermanson, D. J., Hartley, N. D., Gamble-George, J., Brown, N., Shonesy, B. C., Kingsley, P. J., et al. 2013. Substrate-selective COX-2 inhibition decreases anxiety via endocannabinoid activation. *Nature Neuroscience* 16 (9), 1291–1298. doi: 10.1038/nn.3480

Herrera-Solis, A., Vasquez, K. G., and Prospero-Garcia, O. 2010. Acute and subchronic administration of anandamide or oleamide increases REM sleep in rats. *Pharmacology, Biochemistry, and Behavior* 95 (1), 106–112. doi: 10.1016/j.pbb.2009.12.014

Hickman, M., Vickerman, P., Macleod, J., Lewis, G., Zammit, S., Kirkbride, J., et al. 2009. If cannabis caused schizophrenia—how many cannabis users may need to be prevented in order to prevent one case of schizophrenia? England and Wales calculations. *Addiction* 104 (11), 1856–1861. doi: 10.1111/j.1360-0443.2009.02736.x

Hickok, J. T., Roscoe, J. A., Morrow, G. R., King, D. K., Atkins, J. N., and Fitch, T. R. 2003. Nausea and emesis remain significant problems of chemotherapy despite prophylaxis with 5-hydroxytryptamine-3 antiemetics. *Cancer* 97 (11), 2880–2886.

Higgs, S., Barber, D. J., Cooper, A. J., and Terry, P. 2005. Differential effects of two cannabinoid receptor agonists on progressive ratio responding for food and free-feeding in rats. *Behavioural Pharmacology* 16 (5–6), 389–393.

Higuchi, S., Ohji, M., Araki, M., Furuta, R., Katsuki, M., Yamaguchi, R., et al. 2011. Increment of hypothalamic 2-arachidonoylglycerol induces the preference for a high-fat diet via activation of cannabinoid 1 receptors. *Behavioural Brain Research* 216 (1), 477–480. doi: 10.1016/j.bbr.2010.08.042

Hill, A. J., Mercier, M. S., Hill, T. D., Glyn, S. E., Jones, N. A., Yamasaki, Y., et al. 2012. Cannabidivarin is anticonvulsant in mouse and rat. *British Journal of Pharmacology* 167 (8), 1629–1642. doi: 10.1111/j.1476-5381.2012.02207.x

Hill, M. N., Bierer, L. M., Makotkine, I., Golier, J. A., Galea, S., McEwen, B. S., et al. 2013. Reductions in circulating endocannabinoid levels in individuals with post-traumatic stress disorder following exposure to the World Trade Center attacks. *Psychoneuroendocrinology* 38 (12), 2952–2961. doi: 10.1016/j. psyneuen.2013.08.004

Hill, M. N., and Gorzalka, B. B. 2009. Impairments in endocannabinoid signaling and depressive illness. *Journal of the American Medical Association* 301 (11), 1165–1166. doi: 10.1001/jama.2009.369

Hill, M. N., Kumar, S. A., Filipski, S. B., Iverson, M., Stuhr, K. L., Keith, J. M., et al. 2013. Disruption of fatty acid amide hydrolase activity prevents the effects of chronic stress on anxiety and amygdalar microstructure. *Molecular Psychiatry* 18 (10), 1125–1135. doi: 10.1038/mp.2012.90

Hill, M. N., McLaughlin, R. J., Bingham, B., Shrestha, L., Lee, T. T., Gray, J. M., et al. 2010. Endogenous cannabinoid signaling is essential for stress adaptation. *Proceedings of the National Academy of Sciences* 107 (20), 9406–9411. doi: 10.1073/pnas.0914661107

Hill, M. N., Miller, G. E., Ho, W. S., Gorzalka, B. B., and Hillard, C. J. 2008. Serum endocannabinoid content is altered in females with depressive disorders: A preliminary report. *Pharmacopsychiatry* 41 (2), 48–53. doi: 10.1055/s-2007-993211

Hill, M. N., Patel, S., Campolongo, P., Tasker, J. G., Wotjak, C. T., and Bains, J. S. 2010. Functional interactions between stress and the endocannabinoid system: From synaptic signaling to behavioral output. *Journal of Neuroscience* 30 (45), 14980–14986. doi: 10.1523/JNEUROSCI.4283-10.2010

Hill, M. N., and Tasker, J. G. 2012. Endocannabinoid signaling, glucocorticoid-mediated negative feedback, and regulation of the hypothalamic-pituitary-adrenal axis. *Neuroscience* 204, 5–16. doi: 10.1016/j.neuroscience.2011.12.030

Hillard, C. J. 2000. Endocannabinoids and vascular function. *Journal of Pharmacology and Experimental Therapeutics*, 294 (1), 27–32.

Hillard, C. J., Weinlander, K. M., and Stuhr, K. L. 2012. Contributions of endocannabinoid signaling to psychiatric disorders in humans: Genetic and biochemical evidence. *Neuroscience* 204, 207–229. doi: 10.1016/j.neuroscience .2011.11.020

Hilliard, A., Stott, C., Wright, S., Guy, G., Pryce, G., and Al-Izki, S., et al. 2012. Evaluation of the effects of Sativex (THC BDS: CBD BDS) on inhibition of spasticity in a chronic relapsing experimental allergic autoimmune encephalomyelitis: A model of multiple sclerosis. *ISRN Neurology* 802649. doi: 10.5402/2012/802649

Hindocha, C., Freeman, T. P., Schafer, G., Gardener, C., Das, R. K., Morgan, C. J., et al. 2015. Acute effects of delta-9-tetrahydrocannabinol, cannabidiol and their combination on facial emotion recognition: A randomised, double-blind, placebo-controlled study in cannabis users. *European Neuropsychopharmacology* 25 (3), 325–334. doi: 10.1016/j.euroneuro.2014.11.014

Hirvonen, J., Goodwin, R. S., Li, C. T., Terry, G. E., Zoghbi, S. S., Morse, C., et al. 2012. Reversible and regionally selective downregulation of brain cannabinoid CB1 receptors in chronic daily cannabis smokers. *Molecular Psychiatry* 17 (6), 642–649. doi: 10.1038/mp.2011.82

Ho, B. C., Wassink, T. H., Ziebell, S., and Andreasen, N. C. 2011. Cannabinoid receptor 1 gene polymorphisms and marijuana misuse interactions on white matter and cognitive deficits in schizophrenia. *Schizophrenia Research* 128 (1–3), 66–75. doi: 10.1016/j.schres.2011.02.021

Hohmann, A. G., Suplita, R. L., Bolton, N. M., Neely, M. H., Fegley, D., Mangieri, R., et al. 2005. An endocannabinoid mechanism for stress-induced analgesia. *Nature* 435 (7045), 1108–1112. doi: 10.1038/nature03658

Hollister, L. E. 1971. Actions of various marihuana derivatives in man. *Pharmacological Reviews* 23 (4), 349–357.

Hollister, L. E. 1974. Structure-activity relationships in man of cannabis constituents, and homologs and metabolites of delta9-tetrahydrocannabinol. *Pharmacology* 11 (1), 3–11.

Holter, S. M., Kallnik, M., Wurst, W., Marsicano, G., Lutz, B., and Wotjak, C. T. 2005. Cannabinoid CB1 receptor is dispensable for memory extinction in an appetitively-motivated learning task. *European Journal of Pharmacology* 510 (1–2), 69–74. doi: 10.1016/j.ejphar.2005.01.008

Hornby, P. J. 2001. Central neurocircuitry associated with emesis. *American Journal of Medicine* 111 (supplement 8A), 106S–112S.

Horswill, J. G., Bali, U., Shaaban, S., Keily, J. F., Jeevaratnam, P., Babbs, A. J., et al. 2007. PSNCBAM-1, a novel allosteric antagonist at cannabinoid CB1 receptors with hypophagic effects in rats. *British Journal of Pharmacology* 152 (5), 805–814. doi: 10.1038/sj.bjp.0707347

Howlett, A. C., Barth, F., Bonner, T. I., Cabral, G., Casellas, P., Devane, W. A., et al. 2002. International Union of Pharmacology. XXVII. Classification of cannabinoid receptors. *Pharmacological Reviews* 54 (2), 161–202.

Howlett, A. C., Johnson, M. R., Melvin, L. S., and Milne, G. M. 1988. Nonclassical cannabinoid analgetics inhibit adenylate cyclase: Development of a cannabinoid receptor model. *Molecular Pharmacology* 33 (3), 297–302.

Howlett, A. C., Qualy, J. M., and Khachatrian, L. L. 1986. Involvement of Gi in the inhibition of adenylate cyclase by cannabimimetic drugs. *Molecular Pharmacology* 29 (3), 307–313.

Hoyte, C. O., Jacob, J., Monte, A. A., Al-Jumaan, M., Bronstein, A. C., and Heard, K. J. 2012. A characterization of synthetic cannabinoid exposures reported to the National Poison Data System in 2010. *Annals of Emergency Medicine* 60 (4), 435–438. doi: 10.1016/j.annemergmed.2012.03.007

Huestis, M. A. (2005). Pharmacokinetics and metabolism of the plant cannabinoids, delta9-tetrahydrocannabinol, cannabidiol and cannabinol. *Handbook of Experimental Pharmacology*, 168, 657–690.

Huestis, M. A., and Smith, M. L. 2014. Cannabinoid pharmacokinetics and disposition in alternative matrices. In *Handbook of Cannabis*, ed. R. G. Pertwee. Oxford University Press.

Huggins, J. P., Smart, T. S., Langman, S., Taylor, L., and Young, T. 2012. An efficient randomised, placebo-controlled clinical trial with the irreversible fatty acid amide hydrolase-1 inhibitor PF-04457845, which modulates endocannabinoids but fails to induce effective analgesia in patients with pain due to osteoarthritis of the knee. *Pain* 153 (9), 1837–1846. doi: 10.1016/j. pain.2012.04.020

Ignatowska-Jankowska, B. M., Baillie, G. L., Kinsey, S., Crowe, M., Ghosh, S., Owens, R. A., et al. 2015. A Cannabinoid CB1 Receptor-Positive Allosteric Modulator Reduces Neuropathic Pain in the Mouse with No Psychoactive Effects. *Neuropsychopharmacology* 40 (13), 2948–2959. doi: 10.1038/npp.2015.148

Ignatowska-Jankowska, B. M., Ghosh, S., Crowe, M. S., Kinsey, S. G., Niphakis, M. J., Abdullah, R. A., et al. 2014. In vivo characterization of the highly selective monoacylglycerol lipase inhibitor KML29: Antinociceptive activity without cannabimimetic side effects. *British Journal of Pharmacology* 171 (6), 1392–1407. doi: 10.1111/bph.12298

Ishiguro, H., Onaivi, E. S., Horiuchi, Y., Imai, K., Komaki, G., Ishikawa, T., et al. 2011. Functional polymorphism in the GPR55 gene is associated with anorexia nervosa. *Synapse* 65 (2), 103–108. doi: 10.1002/syn.20821

Iskedjian, M., Bereza, B., Gordon, A., Piwko, C., and Einarson, T. R. 2007. Meta-analysis of cannabis based treatments for neuropathic and multiple sclerosis-related pain. *Current Medical Research and Opinion* 23 (1), 17–24. doi: 10.1185/030079906X158066

Iversen, L. L. 2008. *The Science of Marijuana*. Oxford University Press.

Izquierdo, I., Orsingher, O. A., and Berardi, A. C. 1973. Effect of cannabidiol and of other cannabis sativa compounds on hippocampal seizure discharges. *Psychopharmacology* 28 (1), 95–102.

Jackson, N. J., Isen, J. D., Khoddam, R., Irons, D., Tuvblad, C., Iacono, W. G., et al. 2016. Impact of adolescent marijuana use on intelligence: Results from two longitudinal twin studies. *Proceedings of the National Academy of Sciences* 113 (5), E500–E508. doi: 10.1073/pnas.1516648113

Jackson, S. L., Pryce, G., Diemel, D. T., and Baker, D. 2005. Cannabinoid receptor null mice are susceptible to neurofiliment damage and capsase 3 activation. *Neuroscience* 134, 261–268.

Jaques, S. C., Kingsbury, A., Henshcke, P., Chomchai, C., Clews, S., Falconer, J., et al. 2014. Cannabis, the pregnant woman and her child: Weeding out the myths. *Journal of Perinatology* 34 (6), 417–424. doi: 10.1038/jp.2013.180

Jarrett, M. M., Limebeer, C. L., and Parker, L. A. 2005. Effect of Delta9-tetrahydrocannabinol on sucrose palatability as measured by the taste reactivity test. *Physiology & Behavior* 86 (4), 475–479.

Jarrett, M. M., Scantlebury, J., and Parker, L. A. 2007. Effect of Delta (9)-tetrahydrocannabinol on quinine palatability and AM251 on sucrose and quinine palatability using the taste reactivity test. *Physiology & Behavior* 90 (2–3), 425–430. doi: 10.1016/j.physbeh.2006.10.003

Jarzimski, C., Karst, M., Zoerner, A. A., Rakers, C., May, M., Suchy, M. T., et al. 2012. Changes of blood endocannabinoids during anaesthesia: A special case for fatty acid amide hydrolase inhibition by propofol? *British Journal of Clinical Pharmacology* 74 (1), 54–59. doi: 10.1111/j.1365-2125.2012.04175.x

Jayamanne, A., Greenwood, R., Mitchell, V. A., Aslan, S., Piomelli, D., and Vaughan, C. W. 2006. Actions of the FAAH inhibitor URB597 in neuropathic and inflammatory chronic pain models. *British Journal of Pharmacology* 147 (3), 281–288. doi: 10.1038/sj.bjp.0706510

Jenniches, I., Ternes, S., Albayram, O., Otte, D. M., Bach, K., Bindila, L., et al. 2016. Anxiety, stress, and fear response in mice with reduced endocannabinoid levels. *Biological Psychiatry* 79 (10), 858–868. doi: 10.1016/j.biopsych .2015.03.033

Jetly, R., Heber, A., Fraser, G., and Boisvert, D. 2015. The efficacy of nabilone, a synthetic cannabinoid, in the treatment of PTSD-associated nightmares: A preliminary randomized, double-blind, placebo-controlled cross-over design study. *Psychoneuroendocrinology* 51, 585–588. doi: 10.1016/j.psyneuen.2014 .11.002

Jin, K. L., Mao, X. O., Goldsmith, P. C., and Greenberg, D. A. 2000. CB1 cannabinoid receptor induction in experimental stroke. *Annals of Neurology* 48 (2), 257–261.

Jones, N. A., Hill, A. J., Smith, I., Bevan, S. A., Williams, C. M., Whalley, B. J., et al. 2010. Cannabidiol displays antiepileptiform and antiseizure properties in vitro and in vivo. *Journal of Pharmacology and Experimental Therapeutics* 332 (2), 569–577. doi: 10.1124/jpet.109.159145

Joshi, M., Joshi, A., and Bartter, T. 2014. Marijuana and lung diseases. *Current Opinion in Pulmonary Medicine* 20 (2), 173–179. doi: 10.1097/MCP .0000000000000026

Jung, K. M., Clapper, J. R., Fu, J., D'Agostino, G., Guijarro, A., Thongkham, D., et al. 2012. 2-arachidonoylglycerol signaling in forebrain regulates systemic energy metabolism. *Cell Metabolism* 15 (3), 299–310. doi: 10.1016/j.cmet .2012.01.021

Justinova, Z., Ferre, S., Redhi, G. H., Mascia, P., Stroik, J., Quarta, D., et al. 2011. Reinforcing and neurochemical effects of cannabinoid CB1 receptor agonists, but not cocaine, are altered by an adenosine A2A receptor antagonist. *Addiction Biology* 16 (3), 405–415. doi: 10.1111/j.1369-1600.2010.00258.x

Justinova, Z., Mangieri, R. A., Bortolato, M., Chefer, S. I., Mukhin, A. G., Clapper, J. R., et al. 2008. Fatty acid amide hydrolase inhibition heightens anandamide signaling without producing reinforcing effects in primates. *Biological Psychiatry* 64 (11), 930–937. doi: 10.1016/j.biopsych.2008.08.008

Justinova, Z., Panlilio, L. V., Moreno-Sanz, G., Redhi, G. H., Auber, A., Secci, M. E., et al. 2015. Effects of Fatty Acid Amide Hydrolase (FAAH) Inhibitors in Non-Human Primate Models of Nicotine Reward and Relapse. *Neuropsychopharmacology* 40 (9), 2185–2197. doi: 10.1038/npp.2015.62

Kaczocha, M., Glaser, S. T., and Deutsch, D. G. 2009. Identification of intracellular carriers for the endocannabinoid anandamide. *Proceedings of the National Academy of Sciences* 106 (15), 6375–6380. doi: 10.1073/pnas.0901515106

Kamprath, K., Romo-Parra, H., Haring, M., Gaburro, S., Doengi, M., Lutz, B., et al. 2011. Short-term adaptation of conditioned fear responses through endocannabinoid signaling in the central amygdala. *Neuropsychopharmacology* 36 (3), 652–663. doi: 10.1038/npp.2010.196

Karler, R., Cely, W., and Turkanis, S. A. 1973. The anticonvulsant activity of cannabidiol and cannabinol. *Life Sciences* 13 (11), 1527–1531.

Karler, R., and Turkanis, S. A. 1978. Cannabis and epilepsy. *Advances in the Biosciences* 22–23, 619–641.

Karniol, I. G., Shirakawa, I., Kasinski, N., Pfeferman, A., and Carlini, E. A. 1974. Cannabidiol interferes with the effects of delta 9-tetrahydrocannabinol in man. *European Journal of Pharmacology* 28 (1), 172–177.

Karschner, E. L., Darwin, W. D., Goodwin, R. S., Wright, S., and Huestis, M. A. 2011. Plasma cannabinoid pharmacokinetics following controlled oral delta9-tetrahydrocannabinol and oromucosal cannabis extract administration. *Clinical Chemistry* 57 (1), 66–75. doi: 10.1373/clinchem.2010.152439

Karschner, E. L., Darwin, W. D., McMahon, R. P., Liu, F., Wright, S., Goodwin, R. S., et al. 2011. Subjective and physiological effects after controlled Sativex and oral THC administration. *Clinical Pharmacology and Therapeutics* 89 (3), 400–407. doi: 10.1038/clpt.2010.318

Kathuria, S., Gaetani, S., Fegley, D., Valino, F., Duranti, A., Tontini, A., et al. 2003. Modulation of anxiety through blockade of anandamide hydrolysis. *Nature Medicine* 9 (1), 76–81. doi: 10.1038/nm803

Katona, I. 2015. Cannabis and endocannabinoid signaling in epilepsy. In R. G. Pertwee (Ed.), *Endocannabinoids*. Springer.

Katona, I., and Freund, T. F. 2012. Multiple functions of endocannabinoid signaling in the brain. *Annual Review of Neuroscience* 35, 529–558. doi: 10.1146/annurev-neuro-062111-150420

Kay, S. R., Opler, L. A., and Lindenmayer, J. P. 1988. Reliability and validity of the positive and negative syndrome scale for schizophrenics. *Psychiatry Research* 23 (1), 99–110.

Keeler, M. H., and Riefler, C. B. 1967. Grand mal convulsions subsequent to marijuana use. *Diseases of the Nervous System* 28, 474–475.

Kenakin, T. 2013. Allosteric drugs and seven transmembrane receptors. *Current Topics in Medicinal Chemistry* 13 (1), 5–13.

Kiefer, S. W., and Orr, M. R. 1992. Taste avoidance, but not aversion, learning in rats lacking gustatory cortex. *Behavioral Neuroscience* 106 (1), 140–146.

Kim-Cohen, J., Caspi, A., Moffitt, T. E., Harrington, H., Milne, B. J., and Poulton, R. 2003. Prior juvenile diagnoses in adults with mental disorder: Developmental follow-back of a prospective-longitudinal cohort. *Archives of General Psychiatry* 60 (7), 709–717. doi: 10.1001/archpsyc.60.7.709

Kinsey, S. G., Long, J. Z., O'Neal, S. T., Abdullah, R. A., Poklis, J. L., Boger, D. L., et al. 2009. Blockade of endocannabinoid-degrading enzymes attenuates neuropathic pain. *Journal of Pharmacology and Experimental Therapeutics* 330 (3), 902–910. doi: 10.1124/jpet.109.155465

Kinsey, S. G., O'Neal, S. T., Long, J. Z., Cravatt, B. F., and Lichtman, A. H. 2011. Inhibition of endocannabinoid catabolic enzymes elicits anxiolytic-like effects in the marble burying assay. *Pharmacology, Biochemistry, and Behavior* 98 (1), 21–27. doi: 10.1016/j.pbb.2010.12.002

Kinsey, S. G., Wise, L. E., Ramesh, D., Abdullah, R., Selley, D. E., Cravatt, B. F., et al. 2013. Repeated low-dose administration of the monoacylglycerol lipase inhibitor JZL184 retains cannabinoid receptor type 1-mediated antinociceptive and gastroprotective effects. *Journal of Pharmacology and Experimental Therapeutics* 345 (3), 492–501. doi: 10.1124/jpet.112.201426

Kirkham, T. C. 1991. Opioids and feeding reward. *Appetite* 17 (1), 74–75.

Kirkham, T. C., and Williams, C. M. 2001. Synergistic efects of opioid and cannabinoid antagonists on food intake. *Psychopharmacology* 153 (2), 267–270.

Kirkham, T. C., Williams, C. M., Fezza, F., and Di Marzo, V. 2002. Endocannabinoid levels in rat limbic forebrain and hypothalamus in relation to fasting, feeding and satiation: Stimulation of eating by 2-arachidonoyl glycerol. *British Journal of Pharmacology* 136 (4), 550–557. doi: 10.1038/sj.bjp.0704767

Koch, J. E. 2001. Delta (9)-THC stimulates food intake in Lewis rats: Effects on chow, high-fat and sweet high-fat diets. *Pharmacology, Biochemistry, and Behavior* 68 (3), 539–543.

Koethe, D., Giuffrida, A., Schreiber, D., Hellmich, M., Schultze-Lutter, F., Ruhrmann, S., et al. 2009. Anandamide elevation in cerebrospinal fluid in initial prodromal states of psychosis. *British Journal of Psychiatry* 194 (4), 371–372. doi: 10.1192/bjp.bp.108.053843

Koethe, D., Schreiber, D., Giuffrida, A., Mauss, C., Faulhaber, J., Heydenreich, B., et al. 2009. Sleep deprivation increases oleoylethanolamide in human cerebrospinal fluid. *Journal of Neural Transmission* 116 (3), 301–305. doi: 10.1007/s00702-008-0169-6

Kolb, B., Gorny, G., Limebeer, C. L., and Parker, L. A. 2006. Chronic treatment with Delta-9-tetrahydrocannabinol alters the structure of neurons in the nucleus accumbens shell and medial prefrontal cortex of rats. *Synapse* 60 (6), 429–436.

Koppell, B. S., Brust, J. C., Fife, T., Bronstein, J., Youssof, S., Gronseth, G., et al. 2014. Sytematic review: efficacy and safety of medical marijuana in selected neurological disorders: report of the guideline development subcommittee of the american academy of neurology. *Neurology* 82 (17), 1556–1563.

Kumar, G., Stendall, C., Mistry, R., Gurusamy, K., and Walker, D. 2014. A comparison of total intravenous anaesthesia using propofol with sevoflurane or desflurane in ambulatory surgery: Systematic review and meta-analysis. *Anaesthesia* 69 (10), 1138–1150. doi: 10.1111/anae.12713

Kurtzke, J. F. 1993. Epidemiologic evidence for multiple sclerosis as an infection. *Clinical Microbiology Reviews* 6 (4), 382–427.

Kwiatkowska, M., Parker, L. A., Burton, P., and Mechoulam, R. 2004. A comparative analysis of the potential of cannabinoids and ondansetron to suppress cisplatin-induced emesis in the Suncus murinus (house musk shrew). *Psychopharmacology* 174 (2), 254–259.

Labrecque, G., Halle, S., Berthiaume, A., Morin, G., and Morin, P. J. 1978. Potentiation of the epileptogenic effect of penicillin G by marihuana smoking. *Canadian Journal of Physiology and Pharmacology* 56 (1), 87–96.

Langford, R. M., Mares, J., Novotna, A., Vachova, M., Novakova, I., Notcutt, W., et al. 2013. A double-blind, randomized, placebo-controlled, parallel-group study of THC/CBD oromucosal spray in combination with the existing treatment regimen, in the relief of central neuropathic pain in patients with multiple sclerosis. *Journal of Neurology* 260 (4), 984–997. doi: 10.1007/s00415-012-6739-4

Laprairie, R. B., Bagher, A. M., Kelly, M. E., and Denovan-Wright, E. M. 2016. Biased Type 1 cannabinoid receptor signaling influences neuronal viability in a cell culture model of Huntington Disease. *Molecular Pharmacology* 89 (3), 364–375. doi: 10.1124/mol.115.101980

Laprairie, R. B., Kulkarni, A. R., Kulkarni, P. M., Hurst, D. P., Lynch, D., Reggio, P. H., et al. 2016. Mapping cannabinoid 1 receptor allosteric site(s): Critical molecular determinant and signaling profile of GAT100, a novel, potent, and irreversibly binding probe. *ACS Chemical Neuroscience*. doi: 10.1021/acschemneuro.6b00041

Large, M., Sharma, S., Compton, M. T., Slade, T., and Nielssen, O. 2011. Cannabis use and earlier onset of psychosis: A systematic meta-analysis. *Archives of General Psychiatry* 68 (6), 555–561. doi: 10.1001/archgenpsychiatry.2011.5

Lastres-Becker, I., Cebeira, M., de Ceballos, M. L., Zeng, B. Y., Jenner, P., Ramos, J. A., et al. 2001. Increased cannabinoid CB1 receptor binding and activation of GTP-binding proteins in the basal ganglia of patients with Parkinson's syndrome and of MPTP-treated marmosets. *European Journal of Neuroscience* 14 (11), 1827–1832.

Lastres-Becker, I., Molina-Holgado, F., Ramos, J. A., Mechoulam, R., and Fernandez-Ruiz, J. 2005. Cannabinoids provide neuroprotection against 6-hydroxydopamine toxicity in vivo and in vitro: Relevance to Parkinson's disease. *Neurobiology of Disease* 19 (1–2), 96–107. doi: 10.1016/j.nbd.2004.11.009

Lazenka, M. F., Selley, D. E., and Sim-Selley, L. J. 2013. Brain regional differences in CB1 receptor adaptation and regulation of transcription. *Life Sciences* 92 (8–9), 446–452. doi: 10.1016/j.lfs.2012.08.023

Lee, M. C., Ploner, M., Wiech, K., Bingel, U., Wanigasekera, V., Brooks, J., et al. 2013. Amygdala activity contributes to the dissociative effect of cannabis on pain perception. *Pain*, 154 (1), 124–134. doi: 10.1016/j.pain.2012.09.017

Lepore, M., Vorel, S. R., Lowinson, J., and Gardiner, E. L. 1995. Conditioned place preference induced by delta-9-tetrahydrocannabinol: Comparison with cocaine, morphine and food reward. *Life Sciences* 56, 2073–2080.

Leung, K., Elmes, M. W., Glaser, S. T., Deutsch, D. G., & Kaczocha, M. (2013). Role of FAAH-like anandamide transporter in anandamide inactivation. *PLoS One*, 8, e799355.

Leweke, F. M., Giuffrida, A., Koethe, D., Schreiber, D., Nolden, B. M., Kranaster, L., et al. 2007. Anandamide levels in cerebrospinal fluid of first-episode schizophrenic patients: Impact of cannabis use. *Schizophrenia Research* 94 (1–3), 29–36. doi: 10.1016/j.schres.2007.04.025

Leweke, F. M., Piomelli, D., Pahlisch, F., Muhl, D., Gerth, C. W., Hoyer, C., et al. 2012. Cannabidiol enhances anandamide signaling and alleviates psychotic symptoms of schizophrenia. *Translational Psychiatry* 2, e94. doi: 10.1038/tp.2012.15

Li, G. L., Winter, H., Arends, R., Jay, G. W., Le, V., Young, T., et al. 2012. Assessment of the pharmacology and tolerability of PF-04457845, an irreversible inhibitor of fatty acid amide hydrolase-1, in healthy subjects. *British Journal of Clinical Pharmacology* 73 (5), 706–716. doi: 10.1111/j.1365-2125.2011.04137.x

Li, M. C., Brady, J. E., DiMaggio, C. J., Lusardi, A. R., Tzong, K. Y., and Li, G. 2012. Marijuana use and motor vehicle crashes. *Epidemiologic Reviews* 34, 65–72. doi: 10.1093/epirev/mxr017

Lichtman, A. H., Leung, D., Shelton, C. C., Saghatelian, A., Hardouin, C., Boger, D. L., et al. 2004. Reversible inhibitors of fatty acid amide hydrolase that promote analgesia: Evidence for an unprecedented combination of potency and selectivity. *Journal of Pharmacology and Experimental Therapeutics* 311 (2), 441–448. doi: 10.1124/jpet.104.069401

Lichtman, A. H., and Martin, B. R. 1996. Delta 9-tetrahydrocannabinol impairs spatial memory through a cannabinoid receptor mechanism. *Psychopharmacology* 126 (2), 125–131.

Lichtman, A. H., Shelton, C. C., Advani, T., and Cravatt, B. F. 2004. Mice lacking fatty acid amide hydrolase exhibit a cannabinoid receptor-mediated phenotypic hypoalgesia. *Pain* 109 (3), 319–327. doi: 10.1016/j.pain.2004.01.022

Ligresti, A., Moriello, A. S., Starowicz, K., Matias, I., Pisanti, S., De Petrocellis, L., et al. 2006. Antitumor activity of plant cannabinoids with emphasis on the effect of cannabidiol on human breast carcinoma. *Journal of Pharmacology and Experimental Therapeutics* 318 (3), 1375–1387. doi: 10.1124/jpet.106.105247

Limebeer, C. L., Abdullah, R. A., Rock, E. M., Imhof, E., Wang, K., Lichtman, A. H., et al. 2014. Attenuation of anticipatory nausea in a rat model of contextually elicited conditioned gaping by enhancement of the endocannabinoid system. *Psychopharmacology* 231 (3), 603–612.

Limebeer, C. L., Parker, L. A., and Fletcher, P. J. 2004. 5,7-dihydroxytryptamine lesions of the dorsal and median raphe nuclei interfere with lithium-induced conditioned gaping, but not conditioned taste avoidance, in rats. *Behavioral Neuroscience* 118 (6), 1391–1399.

Limebeer, C. L., Rock, E. M., Mechoulam, R., and Parker, L. A. 2012. The antinausea effects of CB1 agonists are mediated by an action at the visceral insular cortex. *British Journal of Pharmacology* 167 (5), 1126–1136.

Limebeer, C. L., Rock, E. M., Puvanenthirarajah, N., Niphakis, M. J., Cravatt, B. F., and Parker, L. A. 2016. Elevation of 2-AG by monoacylglycerol lipase inhibition in the visceral insular cortex interferes with anticipatory nausea in a rat model. *Behavioral Neuroscience* 130 (2), 261–266. doi: 10.1037/bne0000132

Limebeer, C. L., Vemuri, V. K., Bedard, H., Lang, S. T., Ossenkopp, K. P., Makriyannis, A., et al. 2010. Inverse agonism of cannabinoid CB1 receptors potentiates LiCl-induced nausea in the conditioned gaping model in rats. *British Journal of Pharmacology* 161 (2), 336–349. doi: 10.1111/j.1476 -5381.2010.00885.x

Linszen, D. H., Dingemans, P. M., & Lenior, M. E. (1994). Cannabis use and the course of recent-onset schizophrenic disorders. *Archives of General Psychiatry*, 51(4), 273–279.

Liu, B., Song, S., Jones, P. M., and Persaud, S. J. 2015. GPR55: From orphan to metabolic regulator? *Pharmacology & Therapeutics* 145, 35–42. doi: 10.1016/ j.pharmthera.2014.06.007

Llorente-Berzal, A., Terzian, A. L., di Marzo, V., Micale, V., Viveros, M. P., and Wotjak, C. T. 2015. 2-AG promotes the expression of conditioned fear via cannabinoid receptor type 1 on GABAergic neurons. *Psychopharmacology* 232 (15), 2811–2825. doi: 10.1007/s00213-015-3917-y

Long, J. Z., Li, W., Booker, L., Burston, J. J., Kinsey, S. G., Schlosburg, J. E., et al. 2009. Selective blockade of 2-arachidonoylglycerol hydrolysis produces cannabinoid behavioral effects. *Nature Chemical Biology* 5 (1), 37–44.

Long, J. Z., Nomura, D. K., and Cravatt, B. F. 2009. Characterization of monoacylglycerol lipase inhibition reveals differences in central and peripheral endocannabinoid metabolism. *Chemistry & Biology* 16 (7), 744–753.

Ludanyi, A., Eross, L., Czirjak, S., Vajda, J., Halasz, P., Watanabe, M., et al. 2008. Downregulation of the CB1 cannabinoid receptor and related molecular elements of the endocannabinoid system in epileptic human hippocampus. *Journal of Neuroscience* 28 (12), 2976–2990. doi: 10.1523/JNEUROSCI.4465-07.2008

Lupica, C. R., Riegel, A. C., and Hoffman, A. F. 2004. Marijuana and cannabinoid regulation of brain reward circuits. *British Journal of Pharmacology* 143 (2), 227–234. doi: 10.1038/sj.bjp.0705931

Lutz, B., Marsicano, G., Maldonado, R., and Hillard, C. J. 2015. The endocannabinoid system in guarding against fear, anxiety and stress. *Nature Reviews Neuroscience* 16 (12), 705–718. doi: 10.1038/nrn4036

Lynch, M. E., and Campbell, F. 2011. Cannabinoids for treatment of chronic non-cancer pain: a systematic review of randomized trials. *British Journal of Clinical Pharmacology* 72 (5), 735–744. doi: 10.1111/j.1365-2125.2011.03970.x

Lynch, M. E., & Ware, M. A. (2015). Cannabinoids for the treatment of chronic non-cancer pain: An updated systematic review of randomized controlled trials. *Journal of Neuroimmune Pharmacology*, 10, 293–301.

Lynskey, M. T., Glowinski, A. L., Todorov, A. A., Bucholz, K. K., Madden, P. A., Nelson, E. C., et al. 2004. Major depressive disorder, suicidal ideation, and suicide attempt in twins discordant for cannabis dependence and early-onset cannabis use. *Archives of General Psychiatry* 61 (10), 1026–1032. doi: 10.1001/archpsyc.61.10.1026

Ma, L., Wang, L., Yang, F., Meng, X. D., Wu, C., Ma, H., et al. 2014. Disease-modifying effects of RHC80267 and JZL184 in a pilocarpine mouse model of temporal lobe epilepsy. *CNS Neuroscience & Therapeutics* 20 (10), 905–915. doi: 10.1111/cns.12302

Mahler, S. V., Smith, K. S., and Berridge, K. C. 2007. Endocannabinoid hedonic hotspot for sensory pleasure: Anandamide in nucleus accumbens shell enhances 'liking' of a sweet reward. *Neuropsychopharmacology* 32 (11), 2267–2278.

Maldonado, R., Valverde, O., and Berrendero, F. 2006. Involvement of the endocannabinoid system in drug addiction. *Trends in Neurosciences* 29 (4), 225–232. doi: 10.1016/j.tins.2006.01.008

Malfait, A. M., Gallily, R., Sumariwalla, P. F., Malik, A. S., Andreakos, E., Mechoulam, R., et al. 2000. The nonpsychoactive cannabis constituent cannabidiol is an oral anti-arthritic therapeutic in murine collagen-induced arthritis. *Proceedings of the National Academy of Sciences* 97 (17), 9561–9566. doi: 10.1073/pnas.160105897

Manwell, L. A., Satvat, E., Lang, S. T., Allen, C. P., Leri, F., and Parker, L. A. 2009. FAAH inhibitor, URB-597, promotes extinction and CB (1) antagonist, SR141716, inhibits extinction of conditioned aversion produced by naloxone-precipitated morphine withdrawal, but not extinction of conditioned preference produced by morphine in rats. *Pharmacology, Biochemistry, and Behavior* 94 (1), 154–162. doi: 10.1016/j.pbb.2009.08.002

Marrs, W. R., Blankman, J. L., Horne, E. A., Thomazeau, A., Lin, Y. H., Coy, J., et al. 2010. The serine hydrolase ABHD6 controls the accumulation and efficacy of 2-AG at cannabinoid receptors. *Nature Neuroscience* 13 (8), 951–957. doi: 10.1038/nn.2601

Marshall, C. R. 1897. A contribution to the pharmacology of Cannabis indica. *Proceedings of the Cambridge Philosophical Society*, 149–150.

Marsicano, G., Goodenough, S., Monory, K., Hermann, H., Eder, M., Cannich, A., et al. 2003. CB1 cannabinoid receptors and on-demand defense against excitotoxicity. *Science* 302 (5642), 84–88. doi: 10.1126/science.1088208

Marsicano, G., and Lafenetre, P. 2009. Roles of the endocannabinoid system in learning and memory. *Current Topics in Behavioral Neurosciences* 1, 201–230. doi: 10.1007/978-3-540-88955-7_8

Marsicano, G., Wotjak, C. T., Azad, S. C., Bisogno, T., Rammes, G., Cascio, M. G., et al. 2002. The endogenous cannabinoid system controls extinction of aversive memories. *Nature* 418 (6897), 530–534. doi: 10.1038/nature00839

Martin, B. R., Compton, D. R., Thomas, B. F., Prescott, W. R., Little, P. J., Razdan, R. K., et al. 1991. Behavioral, biochemical, and molecular modeling evaluations of cannabinoid analogs. *Pharmacology, Biochemistry, and Behavior* 40 (3), 471–478.

Martin-Santos, R., Crippa, J. A., Batalla, A., Bhattacharyya, S., Atakan, Z., Borgwardt, S., et al. 2012. Acute effects of a single, oral dose of d9-tetrahydrocannabinol (THC) and cannabidiol (CBD) administration in healthy volunteers. *Current Pharmaceutical Design* 18 (32), 4966–4979.

Mascia, P., Pistis, M., Justinova, Z., Panlilio, L. V., Luchicchi, A., Lecca, S., et al. 2011. Blockade of nicotine reward and reinstatement by activation of alpha-type peroxisome proliferator-activated receptors. *Biological Psychiatry* 69 (7), 633–641. doi: 10.1016/j.biopsych.2010.07.009

Matias, I., Gatta-Cherifi, B., Tabarin, A., Clark, S., Leste-Lasserre, T., Marsicano, G., et al. 2012. Endocannabinoids measurement in human saliva as potential biomarker of obesity. *PLoS One* 7 (7), e42399. doi: 10.1371/journal.pone.0042399

Mazzola, C., Medalie, J., Scherma, M., Panlilio, L. V., Solinas, M., Tanda, G., et al. 2009. Fatty acid amide hydrolase (FAAH) inhibition enhances memory acquisition through activation of PPAR-alpha nuclear receptors. *Learning & Memory* 16 (5), 332–337. doi: 10.1101/lm.1145209

McEwen, B. S. 2007. Physiology and neurobiology of stress and adaptation: Central role of the brain. *Physiological Reviews* 87 (3), 873–904. doi: 10.1152/physrev.00041.2006

McGrath, J., Welham, J., Scott, J., Varghese, D., Degenhardt, L., Hayatbakhsh, M. R., et al. 2010. Association between cannabis use and psychosis-related outcomes using sibling pair analysis in a cohort of young adults. *Archives of General Psychiatry* 67 (5), 440–447. doi: 10.1001/archgenpsychiatry.2010.6

McLaughlin, P. J., Winston, K. M., Limebeer, C. L., Parker, L. A., Makriyannis, A., and Salamone, J. D. 2005. The cannabinoid CB1 antagonist AM 251 produces food avoidance and behaviors associated with nausea but does not impair feeding efficiency in rats. *Psychopharmacology* 180 (2), 286–293.

Mead, A. P. 2014. International Control of Cannabis. In *Handbook of Cannabis*, ed. R. G. Pertwee. Oxford University Press.

Mechoulam, R. 2002. Discovery of endocannabinoids and some random thoughts on their possible roles in neuroprotection and aggression. *Prostaglandins, Leukotrienes, and Essential Fatty Acids* 66 (2–3), 93–99. doi: 10.1054/plef.2001.0340

Mechoulam, R. 2005. Plant cannabinoids: A neglected pharmacological treasure trove. *British Journal of Pharmacology* 146 (7), 913–915. doi: 10.1038/sj.bjp.0706415

Mechoulam, R., Ben-Shabat, S., Hanus, L., Ligumsky, M., Kaminski, N. E., Schatz, A. R., et al. 1995. Identification of an endogenous 2-monoglyceride, present in canine gut, that binds to cannabinoid receptors. *Biochemical Pharmacology* 50 (1), 83–90.

Mechoulam, R., & Carlini, E. A. (1978). Toward drugs derived from cannabis. *Naturwissenschaften*, 65(4), 174–179.

Mechoulam, R., Feigenbaum, J. J., Lander, N., Segal, M., Jarbe, T. U., Hiltunen, A. J., et al. 1988. Enantiomeric cannabinoids: Stereospecificity of psychotropic activity. *Experientia* 44 (9), 762–764.

Mechoulam, R., and Gaoni, Y. 1967. The absolute configuration of delta-1-tetrahydrocannabinol, the major active constituent of hashish. *Tetrahedron Letters* 12, 1109–1111.

Mechoulam, R., and Hanus, L. 2000. A historical overview of chemical research on cannabinoids. *Chemistry and Physics of Lipids* 108 (1–2), 1–13.

Mechoulam, R., Hanus, L. O., Pertwee, R., and Howlett, A. C. 2014. Early phytocannabinoid chemistry to endocannabinoids and beyond. *Nature Reviews Neuroscience* 15 (11), 757–764. doi: 10.1038/nrn3811

Mechoulam, R., Parker, L., and Gallily, R. 2002. Cannabidiol: An overview of some pharmacological aspects. *Journal of Clinical Pharmacology* 42 (11), 11S–19S. doi: 10.1177/0091270002238789

Mechoulam, R., and Parker, L. A. 2013. The endocannabinoid system and the brain. *Annual Review of Psychology* 64, 21–47.

Mechoulam, R., Parker, L. A., and Gallily, R. 2002. Cannabidiol: An overview of some pharmacological aspects. *Journal of Clinical Pharmacology* 42 (11 Suppl), 11S–19S.

Mechoulam, R., Peters, M., Murillo-Rodriguez, E., and Hanus, L. O. 2007. Cannabidiol—recent advances. *Chemistry & Biodiversity* 4 (8), 1678–1692. doi: 10.1002/cbdv.200790147

Mechoulam, R., and Shvo, Y. 1963. Hashish. I. The structure of cannabidiol. *Tetrahedron* 19 (12), 2073–2078.

Meier, M. H., Caspi, A., Ambler, A., Harrington, H., Houts, R., Keefe, R. S., et al. 2012. Persistent cannabis users show neuropsychological decline from childhood to midlife. *Proceedings of the National Academy of Sciences* 109 (40), E2657–E2664. doi: 10.1073/pnas.1206820109

Meiri, E., Jhangiani, H., Vredenburgh, J. J., Barbato, L. M., Carter, F. J., Yang, H. M., et al. 2007. Efficacy of dronabinol alone and in combination with ondansetron versus ondansetron alone for delayed chemotherapy-induced nausea and vomiting. *Current Medical Research and Opinion* 23 (3), 533–543. doi: 10.1185/030079907X167525

Melis, M., Pillolla, G., Luchicchi, A., Muntoni, A. L., Yasar, S., Goldberg, S. R., et al. 2008. Endogenous fatty acid ethanolamides suppress nicotine-induced activation of mesolimbic dopamine neurons through nuclear receptors. *Journal of Neuroscience* 28 (51), 13985–13994. doi: 10.1523/JNEUROSCI.3221-08.2008

Melis, M., and Pistis, M. 2007. Endocannabinoid signaling in midbrain dopamine neurons: More than physiology? *Current Neuropharmacology* 5 (4), 268–277. doi: 10.2174/157015907782793612

Merkus, F. W. H. M. 1971. Cannabivarin and tetrahydrocannabivarin, two new constituents of hashish. *Nature* 232, 579–580.

Meye, F. J., Trezza, V., Vanderschuren, L. J., Ramakers, G. M., and Adan, R. A. 2013. Neutral antagonism at the cannabinoid 1 receptor: A safer treatment for obesity. *Molecular Psychiatry* 18 (12), 1294–1301. doi: 10.1038/mp.2012.145

Midanik, L. T., Tam, T. W., and Weisner, C. 2007. Concurrent and simultaneous drug and alcohol use: Results of the 2000 National Alcohol Survey. *Drug and Alcohol Dependence* 90 (1), 72–80. doi: 10.1016/j.drugalcdep.2007.02.024

Mills, J. H. 2003. *Cannabis Britannica: Empire, Trade, and Prohibition*. Oxford University Press.

Milton, N. G. 2002. Anandamide and noladin ether prevent neurotoxicity of the human amyloid-beta peptide. *Neuroscience Letters* 332 (2), 127–130.

Miner, W. D., and Sanger, G. J. 1986. Inhibition of cisplatin-induced vomiting by selective 5-hydroxytryptamine M-receptor antagonism. *British Journal of Pharmacology* 88 (3), 497–499.

Mokrysz, C., Landy, R., Gage, S. H., Munafo, M. R., Roiser, J. P., and Curran, H. V. 2016. Are IQ and educational outcomes in teenagers related to their cannabis use? A prospective cohort study. *Journal of Psychopharmacology* 30 (2), 159–168. doi: 10.1177/0269881115622241

Monory, K., Massa, F., Egertova, M., Eder, M., Blaudzun, H., Westenbroek, R., et al. 2006. The endocannabinoid system controls key epileptogenic circuits in the hippocampus. *Neuron* 51 (4), 455–466. doi: 10.1016/j.neuron.2006 .07.006

Monteleone, P., and Maj, M. 2013. Dysfunctions of leptin, ghrelin, BDNF and endocannabinoids in eating disorders: Beyond the homeostatic control of food intake. *Psychoneuroendocrinology* 38 (3), 312–330. doi: 10.1016/j.psyneuen .2012.10.021

Monteleone, P., Matias, I., Martiadis, V., De Petrocellis, L., Maj, M., and Di Marzo, V. 2005. Blood levels of the endocannabinoid anandamide are increased in anorexia nervosa and in binge-eating disorder, but not in bulimia nervosa. *Neuropsychopharmacology* 30 (6), 1216–1221. doi: 10.1038/sj.npp.1300695

Monteleone, P., Piscitelli, F., Scognamiglio, P., Monteleone, A. M., Canestrelli, B., Di Marzo, V., et al. 2012. Hedonic eating is associated with increased peripheral levels of ghrelin and the endocannabinoid 2-arachidonoyl-glycerol in healthy humans: A pilot study. *Journal of Clinical Endocrinology and Metabolism* 97 (6), E917–E924. doi: 10.1210/jc.2011-3018

Moore, T. H., Zammit, S., Lingford-Hughes, A., Barnes, T. R., Jones, P. B., Burke, M., et al. 2007. Cannabis use and risk of psychotic or affective mental health outcomes: A systematic review. *Lancet* 370 (9584), 319–328. doi: 10.1016/ S0140-6736(07)61162-3

Morena, M., Patel, S., Bains, J. S., and Hill, M. N. 2016. Neurobiological Interactions Between Stress and the Endocannabinoid System. *Neuropsychopharmacology* 41 (1), 80–102. doi: 10.1038/npp.2015.166

Moreno-Navarrete, J. M., Catalan, V., Whyte, L., Diaz-Arteaga, A., Vazquez-Martinez, R., Rotellar, F., et al. 2012. The L-alpha-lysophosphatidylinositol/GPR55 system and its potential role in human obesity. *Diabetes* 61 (2), 281–291. doi: 10.2337/db11-0649

Morgan, C. J., and Curran, H. V. 2008. Effects of cannabidiol on schizophrenia-like symptoms in people who use cannabis. *British Journal of Psychiatry* 192 (4), 306–307. doi: 10.1192/bjp.bp.107.046649

Morgan, C. J., Freeman, T. P., Schafer, G. L., and Curran, H. V. 2010. Cannabidiol attenuates the appetitive effects of Delta 9-tetrahydrocannabinol in humans smoking their chosen cannabis. *Neuropsychopharmacology* 35 (9), 1879–1885. doi: 10.1038/npp.2010.58

Morgan, C. J., Schafer, G., Freeman, T. P., and Curran, H. V. 2010. Impact of cannabidiol on the acute memory and psychotomimetic effects of smoked cannabis: naturalistic study: naturalistic study [corrected]. *British Journal of Psychiatry* 197 (4), 285–290. doi: 10.1192/bjp.bp.110.077503

Morrow, G. R., and Dobkin, P. L. 1987. Behavioral approaches for the management of adverse side effects of cancer treatment. *Psychiatric Medicine* 5 (4), 299–314.

Moskowitz, M. A. 2010. Brain protection: Maybe yes, maybe no. *Stroke* 41 (10 Suppl), S85–S86. doi: 10.1161/STROKEAHA.110.598458

Muller-Vahl, K. R., Schneider, U., and Emrich, H. M. 1999. Nabilone increases choreatic movements in Huntington's disease. *Movement Disorders* 14 (6), 1038–1040.

Munro, S., Thomas, K. L., and Abu-Shaar, M. 1993. Molecular characterization of a peripheral receptor for cannabinoids. *Nature* 365 (6441), 61–65. doi: 10.1038/365061a0

Murillo-Rodriguez, E., Aguilar-Turton, L., Mijangos-Moreno, S., Sarro-Rimirez, A., and Arias-Carrion, O. 2014. Phytocannabinoids as novel therapeutic agents for sleep. In *Handbook of Cannabis*, ed. R. G. Pertwee. Oxford University Press.

Murillo-Rodriguez, E., Millan-Aldaco, D., Palomero-Rivero, M., Mechoulam, R., and Drucker-Colin, R. 2006. Cannabidiol, a constituent of Cannabis sativa, modulates sleep in rats. *FEBS Letters* 580 (18), 4337–4345. doi: 10.1016/j.febslet.2006.04.102

Murray, R. M., Morrison, P. D., Henquet, C., and Di Forti, M. 2007. Cannabis, the mind and society: The hash realities. *Nature Reviews Neuroscience* 8 (11), 885–895. doi: 10.1038/nrn2253

Naderi, N., Ahmad-Molaei, L., Aziz Ahari, F., and Motamedi, F. 2011. Modulation of anticonvulsant effects of cannabinoid compounds by GABA-A receptor agonist in acute pentylenetetrazole model of seizure in rat. *Neurochemical Research* 36 (8), 1520–1525. doi: 10.1007/s11064-011-0479-1

Naidu, P. S., Booker, L., Cravatt, B. F., and Lichtman, A. H. 2009. Synergy between enzyme inhibitors of fatty acid amide hydrolase and cyclooxygenase in visceral nociception. *Journal of Pharmacology and Experimental Therapeutics* 329 (1), 48–56. doi: 10.1124/jpet.108.143487

Nakamura, E. M., da Silva, E. A., Concilio, G. V., Wilkinson, D. A., and Masur, J. 1991. Reversible effects of acute and long-term administration of delta-9-tetrahydrocannabinol (THC) on memory in the rat. *Drug and Alcohol Dependence* 28 (2), 167–175.

Naqvi, N. H., Gaznick, N., Tranel, D., and Bechara, A. 2014. The insula: A critical neural substrate for craving and drug seeking under conflict and risk. *Annals of the New York Academy of Sciences* 1316, 53–70. doi: 10.1111/nyas.12415

Nashold, B. S., Wilson, W. P., and Slaughter, D. G. 1969. Sensations evoked by stimulation in the midbrain of man. *Journal of Neurosurgery* 30 (1), 14–24. doi: 10.3171/jns.1969.30.1.0014

Nicholson, A. N., Turner, C., Stone, B. M., and Robson, P. J. 2004. Effect of Delta-9-tetrahydrocannabinol and cannabidiol on nocturnal sleep and early-morning behavior in young adults. *Journal of Clinical Psychopharmacology* 24 (3), 305–313.

Niphakis, M. J., Cognetta, A. B., III, Chang, J. W., Buczynski, M. W., Parsons, L. H., Byrne, F., et al. 2013. Evaluation of NHS Carbamates as a Potent and Selective Class of Endocannabinoid Hydrolase Inhibitors. *ACS Chemical Neuroscience* 4 (9), 1322–1332.

Nissen, S. E., Nicholls, S. J., Wolski, K., Rodes-Cabau, J., Cannon, C. P., Deanfield, J. E., et al. 2008. Effect of rimonabant on progression of atherosclerosis in patients with abdominal obesity and coronary artery disease: The STRADIVARIUS randomized controlled trial. *Journal of the American Medical Association* 299 (13), 1547–1560. doi: 10.1001/jama.299.13.1547

Niyuhire, F., Varvel, S. A., Martin, B. R., and Lichtman, A. H. 2007. Exposure to marijuana smoke impairs memory retrieval in mice. *Journal of Pharmacology and Experimental Therapeutics* 322 (3), 1067–1075. doi: 10.1124/jpet.107.119594

Niyuhire, F., Varvel, S. A., Thorpe, A. J., Stokes, R. J., Wiley, J. L., and Lichtman, A. H. 2007. The disruptive effects of the CB1 receptor antagonist rimonabant on extinction learning in mice are task-specific. *Psychopharmacology* 191 (2), 223–231. doi: 10.1007/s00213-006-0650-6

Notcutt, W., Langford, R., Davies, P., Ratcliffe, S., and Potts, R. 2012. A placebo-controlled, parallel-group, randomized withdrawal study of subjects with symptoms of spasticity due to multiple sclerosis who are receiving long-term Sativex (R) (nabiximols). *Multiple Sclerosis* 18 (2), 219–228. doi: 10.1177/1352458511419700

Nunez, E., Benito, C., Pazos, M. R., Barbachano, A., Fajardo, O., Gonzalez, S., et al. 2004. Cannabinoid CB2 receptors are expressed by perivascular microglial cells in the human brain: An immunohistochemical study. *Synapse* 53 (4), 208–213. doi: 10.1002/syn.20050

O'Brien, L. D., Limebeer, C. L., Rock, E. M., Bottegoni, G., Piomelli, D., and Parker, L. A. 2013. Anandamide transport inhibition by ARN272 attenuates nausea-induced behaviour in rats, and vomiting in shrews (Suncus murinus). *British Journal of Pharmacology* 170 (5), 1130–1136. doi: 10.1111/bph.12360

O'Brien, L. D., Sticht, M. A., Mitchnick, K. A., Limebeer, C. L., Parker, L. A., and Winters, B. D. 2014. CB1 receptor antagonism in the granular insular cortex or somatosensory area facilitates consolidation of object recognition memory. *Neuroscience Letters* 578, 192–196. doi: 10.1016/j.neulet.2014.06.056

O'Brien, L. D., Wills, K. L., Segsworth, B., Dashney, B., Rock, E. M., Limebeer, C. L., et al. 2013. Effect of chronic exposure to rimonabant and phytocannabinoids on anxiety-like behavior and saccharin palatability. *Pharmacology, Biochemistry, and Behavior* 103 (3), 597–602. doi: 10.1016/j.pbb.2012.10.008

O'Brien, L. D., Limebeer, C. L., Rock, E. M., Bottegoni, G., Piomelli, D., & Parker, L. A. 2013. Anandamide transport inhibition by ARN272 attenuates nausea-induced behavior in rats and vomiting in shrews (*Suncus murinus*). *British Journal of Pharmacology*, 170, 1130–1136.

Oka, S., Nakajima, K., Yamashita, A., Kishimoto, S., and Sugiura, T. 2007. Identification of GPR55 as a lysophosphatidylinositol receptor. *Biochemical and Biophysical Research Communications* 362 (4), 928–934. doi: 10.1016/j.bbrc.2007.08.078

Oleson, E. B., Beckert, M. V., Morra, J. T., Lansink, C. S., Cachope, R., Abdullah, R. A., et al. 2012. Endocannabinoids shape accumbal encoding of cue-motivated behavior via CB1 receptor activation in the ventral tegmentum. *Neuron* 73 (2), 360–373. doi: 10.1016/j.neuron.2011.11.018

Onaivi, E. S., Ishiguro, H., Gong, J. P., Patel, S., Meozzi, P. A., Myers, L., et al. 2008. Functional expression of brain neuronal CB2 cannabinoid receptors are involved in the effects of drugs of abuse and in depression. *Annals of the New York Academy of Sciences* 1139, 434–449. doi: 10.1196/annals.1432.036

Osei-Hyiaman, D., DePetrillo, M., Pacher, P., Liu, J., Radaeva, S., Batkai, S., et al. 2005. Endocannabinoid activation at hepatic CB1 receptors stimulates fatty acid synthesis and contributes to diet-induced obesity. *Journal of Clinical Investigation* 115 (5), 1298–1305. doi: 10.1172/JCI23057

O'Shaughnessy, W. B. 1839. On the preparations of Indian hemp, or gunja (*Cannabis indica*); their effects ont he animal system in health, and their utility in the treatment of tentanus and other convulsive diseases. *Transactions of the Medical and Physical Society of Bengal*, 421–461.

O'Shaughnessy, W. B. 1840. On the preparations of the Indian hemp, or gunja (*Cannabis indica*). *Transactions of the Medical and Physical Society of Bengal*, 71–102.

O'Shea, M., McGregor, I. S., and Mallet, P. E. 2006. Repeated cannabinoid exposure during perinatal, adolescent or early adult ages produces similar longlasting deficits in object recognition and reduced social interaction in rats. *Journal of Psychopharmacology* 20 (5), 611–621. doi: 10.1177/0269881106065188

Pacher, P., and Kunos, G. 2013. Modulating the endocannabinoid system in human health and disease—successes and failures. *FEBS Journal* 280 (9), 1918–1943. doi: 10.1111/febs.12260

Pacher, P., and Mechoulam, R. 2011. Is lipid signaling through cannabinoid 2 receptors part of a protective system? *Progress in Lipid Research* 50 (2), 193–211. doi: 10.1016/j.plipres.2011.01.001

Paisley, S., Beard, S., Hunn, A., and Wight, J. 2002. Clinical effectiveness of oral treatments for spasticity in multiple sclerosis: A systematic review. *Multiple Sclerosis* 8 (4), 319–329.

Palazuelos, J., Aguado, T., Pazos, M. R., Julien, B., Carrasco, C., Resel, E., et al. 2009. Microglial CB2 cannabinoid receptors are neuroprotective in Huntington's disease excitotoxicity. *Brain* 132 (11), 3152–3164. doi: 10.1093/brain/awp239

Pamplona, F. A., Ferreira, J., Menezes de Lima, O., Jr., Duarte, F. S., Bento, A. F., Forner, S., et al. 2012. Anti-inflammatory lipoxin A4 is an endogenous allosteric enhancer of CB1 cannabinoid receptor. *Proceedings of the National Academy of Sciences* 109 (51), 21134–21139. doi: 10.1073/pnas.1202906109

Pan, B., Wang, W., Zhong, P., Blankman, J. L., Cravatt, B. F., and Liu, Q. S. 2011. Alterations of endocannabinoid signaling, synaptic plasticity, learning, and memory in monoacylglycerol lipase knock-out mice. *Journal of Neuroscience* 31 (38), 13420–13430. doi: 10.1523/JNEUROSCI.2075-11.2011

Panikashvili, D., Simeonidou, C., Ben-Shabat, S., Hanus, L., Breuer, A., Mechoulam, R., et al. 2001. An endogenous cannabinoid (2-AG) is neuroprotective after brain injury. *Nature* 413 (6855), 527–531. doi: 10.1038/35097089

Parker, L. A. 2014. Conditioned flavor avoidance and conditioned gaping: Rat models of conditioned nausea. *European Journal of Pharmacology* 722, 122–133.

Parker, L. A., and Gillies, T. 1995. THC-induced place and taste aversions in Lewis and Sprague-Dawley rats. *Behavioral Neuroscience* 109 (1), 71–78. doi: 10.1037/0735-7044.109.1.71

Parker, L. A., Limebeer, C. L., Rock, E. M., Litt, D. L., Kwiatkowska, M., and Piomelli, D. 2009. The FAAH inhibitor URB-597 interferes with cisplatin- and nicotine-induced vomiting in the Suncus murinus (house musk shrew). *Physiology & Behavior* 97 (1), 121–124. doi: 10.1016/j.physbeh.2009.02.014

Parker, L. A., Limebeer, C. L., Rock, E. M., Sticht, M. A., Ward, J., Turvey, G., Makriyannis, A., et al. 2016. A comparison of novel, selective fatty acid amide hydrolase (FAAH), monoacylglycerol lipase (MAGL) or dual FAAH/MAGL inhibitors to suppress acute and anticipatory nausea in rat models. *Psychopharmacology* 233 (12), 2265–2275. doi: 10.1007/s00213-016-4277-y

Parker, L. A., Mechoulam, R., and Schlievert, C. 2002. Cannabidiol, a non-psychoactive component of cannabis and its synthetic dimethylheptyl homolog suppress nausea in an experimental model with rats. *Neuroreport* 13 (5), 567–570. doi: 10.1097/00001756-200204160-00006

Parker, L. A., Mechoulam, R., Schlievert, C., Abbott, L., Fudge, M. L., and Burton, P. 2003. Effects of cannabinoids on lithium-induced conditioned rejection reactions in a rat model of nausea. *Psychopharmacology* 166 (2), 156–162.

Parker, L. A., Niphakis, M. J., Downey, R., Limebeer, C. L., Rock, E. M., Sticht, M. A., et al. 2015. Effect of selective inhibition of monoacylglycerol lipase (MAGL) on acute nausea, anticipatory nausea, and vomiting in rats and Suncus murinus. *Psychopharmacology* 232 (3), 583–593. doi: 10.1007/s00213-014-3696-x

Parker, L. A., Rock, E. M., and Limebeer, C. L. 2011. Regulation of nausea and vomiting by cannabinoids. *British Journal of Pharmacology* 163 (7), 1411–1422. doi: 10.1111/j.1476-5381.2010.01176.x

Parker, L. A., Rock, E. M., Sticht, M. A., Wills, K. L., and Limebeer, C. L. 2015. Cannabinoids suppress acute and anticipatory nausea in preclinical rat models of conditioned gaping. *Clinical Pharmacology and Therapeutics* 97 (6), 559–561. doi: 10.1002/cpt.98

Parmentier-Batteur, S., Jin, K., Mao, X. O., Xie, L., and Greenberg, D. A. 2002. Increased severity of stroke in CB1 cannabinoid receptor knock-out mice. *Journal of Neuroscience* 22 (22), 9771–9775.

Parolaro, D., Zamberletti, E., and Rubino, T. 2014. Cannabidiol/Phytocannabinoids: A new opportunity for schizophrenia treatment? In *Handbook of Cannabis*, ed. R. G. Pertwee. Oxford University Press.

Parsons, L. H., and Hurd, Y. L. 2015. Endocannabinoid signalling in reward and addiction. *Nature Reviews. Neuroscience* 16 (10), 579–594. doi: 10.1038/nrn4004

Patel, S., Hill, M. N., & Hillard, C. J. 2014. Effects of phytocannabinoids on anxiety, mood, and the endocrine system. In R. G. Pertwee (Ed.), *Handbook of Cannabis*. Oxford University Press.

Patel, S., and Hillard, C. J. 2006. Pharmacological evaluation of cannabinoid receptor ligands in a mouse model of anxiety: Further evidence for an anxiolytic role for endogenous cannabinoid signaling. *Journal of Pharmacology and Experimental Therapeutics* 318 (1), 304–311. doi: 10.1124/jpet.106.101287

Patel, S., Wohlfeil, E. R., Rademacher, D. J., Carrier, E. J., Perry, L. J., Kundu, A., et al. 2003. The general anesthetic propofol increases brain N-arachidonylethanolamine (anandamide) content and inhibits fatty acid amide hydrolase. *British Journal of Pharmacology* 139 (5), 1005–1013. doi: 10.1038/sj.bjp.0705334

Patton, G. C., Coffey, C., Carlin, J. B., Degenhardt, L., Lynskey, M., and Hall, W. 2002. Cannabis use and mental health in young people: Cohort study. *BMJ* 325 (7374), 1195–1198.

Penfield, W., and Faulk, M. E. 1955. The insula; further observations on its function. *Brain* 78 (4), 445–470.

Perez-Rial, S., Garcia-Gutierrez, M. S., Molina, J. A., Perez-Nievas, B. G., Ledent, C., Leiva, C., et al. 2011. Increased vulnerability to 6-hydroxydopamine lesion and reduced development of dyskinesias in mice lacking CB1 cannabinoid receptors. *Neurobiology of Aging* 32 (4), 631–645. doi: 10.1016/j. neurobiolaging.2009.03.017

Pertwee, R. G. 1972. The ring test: A quantitative method for assessing the 'cataleptic' effect of cannabis in mice. *British Journal of Pharmacology* 46 (4), 753–763.

Pertwee, R. G. 2005. The therapeutic potential of drugs that target cannabinoid receptors or modulate the tissue levels or actions of endocannabinoids. *AAPS Journal* 7 (3), E625–E654. doi: 10.1208/aapsj070364

Pertwee, R. G. 2008. The diverse CB1 and CB2 receptor pharmacology of three plant cannabinoids: delta9-tetrahydrocannabinol, cannabidiol and delta9-tetrahydrocannabivarin. *British Journal of Pharmacology* 153 (2), 199–215. doi: 10.1038/sj.bjp.0707442

Pertwee, R. G. 2009. Emerging strategies for exploiting cannabinoid receptor agonists as medicines. *British Journal of Pharmacology* 156 (3), 397–411. doi: 10.1111/j.1476-5381.2008.00048.x

Pertwee, R. G. 2015. Endocannabinoids and their pharmacological actions. *Handbook of Experimental Pharmacology* 231, 1–37.

Pertwee, R. G., Howlett, A. C., Abood, M. E., Alexander, S. P., Di Marzo, V., Elphick, M. R., et al. 2010. International Union of Basic and Clinical Pharmacology. LXXIX. Cannabinoid receptors and their ligands: Beyond CB (1) and CB (2). *Pharmacological Reviews* 62 (4), 588–631. doi: 10.1124/pr.110.003004

Pertwee, R. G., Thomas, A., Stevenson, L. A., Maor, Y., and Mechoulam, R. 2005. Evidence that (-)-7-hydroxy-4'-dimethylheptyl-cannabidiol activates a non-CB (1), non-CB (2), non-TRPV1 target in the mouse vas deferens. *Neuropharmacology* 48 (8), 1139–1146. doi: 10.1016/j.neuropharm.2005.01.010

Pertwee, R. G., Thomas, A., Stevenson, L. A., Ross, R. A., Varvel, S. A., Lichtman, A. H., et al. 2007. The psychoactive plant cannabinoid, Delta9-tetrahydrocannabinol, is antagonized by Delta8- and Delta9-tetrahydrocannabivarin in mice in vivo. *British Journal of Pharmacology* 150 (5), 586–594. doi: 10.1038/sj.bjp.0707124

Piomelli, D., Haney, M., Budney, A. J., and Piazza, P. V. 2016. Legal or illegal, cannabis is still addictive. *Cannabis and Cannabinoid Research* 1 (1), 47–53.

Piomelli, D., & Russo, E. B. (2016). The *Cannabis sativa* versus *Cannabis indica* Debate: An Interview with Ethan Russo, MD. *Cannabis and Cannabinoid Research*, 1(1), 44–46. doi:10.1089/can.2015.29003.ebr.

Pi-Sunyer, F. X., Aronne, L. J., Heshmati, H. M., Devin, J., Rosenstock, J., et al. 2006. Effect of rimonabant, a cannabinoid-1 receptor blocker, on weight and cardiometabolic risk factors in overweight or obese patients. *Journal of the American Medical Association* 295 (7), 761–775. doi: 10.1001/jama.295.7.761

Pope, H. G., Jr. 2002. Cannabis, cognition, and residual confounding. *Journal of the American Medical Association* 287 (9), 1172–1174.

Portenoy, R. K., Ganae-Motan, E. D., Allende, S., Yanagihara, R., Shaiova, L., Weinstein, S., et al. 2012. Nabiximols for opioid-treated cancer patients with poorly-controlled chronic pain: A randomized, placebo-controlled, graded-dose trial. *Journal of Pain* 13 (5), 438–449. doi: 10.1016/j.jpain.2012.01.003

Porter, B. E., and Jacobson, C. 2013. Report of a parent survey of cannabidiol-enriched cannabis use in pediatric treatment-resistant epilepsy. *Epilepsy & Behavior* 29 (3), 574–577. doi: 10.1016/j.yebeh.2013.08.037

Potter, D. 2014. Cannabis horticulture. In *Handbook of Cannabis*, ed. R. G. Pertwee. Oxford University Press.

Power, R. A., Verweij, K. J., Zuhair, M., Montgomery, G. W., Henders, A. K., Heath, A. C., et al. 2014. Genetic predisposition to schizophrenia associated with increased use of cannabis. *Molecular Psychiatry* 19 (11), 1201–1204. doi: 10.1038/mp.2014.51

Press, C. A., Knupp, K. G., and Chapman, K. E. 2015. Parental reporting of response to oral cannabis extracts for treatment of refractory epilepsy. *Epilepsy & Behavior* 45, 49–52. doi: 10.1016/j.yebeh.2015.02.043

Price, D. A., Martinez, A. A., Seillier, A., Koek, W., Acosta, Y., Fernandez, E., et al. 2009. WIN55,212–2, a cannabinoid receptor agonist, protects against nigrostriatal cell loss in the 1-methyl-4-phenyl-1,2,3,6-tetrahydropyridine mouse model of Parkinson's disease. *European Journal of Neuroscience* 29 (11), 2177–2186. doi: 10.1111/j.1460-9568.2009.06764.x

Price, M. R., Baillie, G. L., Thomas, A., Stevenson, L. A., Easson, M., Goodwin, R., et al. 2005. Allosteric modulation of the cannabinoid CB1 receptor. *Molecular Pharmacology* 68 (5), 1484–1495. doi: 10.1124/mol.105.016162

Pryce, G., Ahmed, Z., Hankey, D. J., Jackson, S. J., Croxford, J. L., Pocock, J. M., et al. 2003. Cannabinoids inhibit neurodegeneration in models of multiple sclerosis. *Brain* 126 (10), 2191–2202. doi: 10.1093/brain/awg224

Pryce, G., and Baker, D. 2007. Control of spasticity in a multiple sclerosis model is mediated by CB1, not CB2, cannabinoid receptors. *British Journal of Pharmacology* 150 (4), 519–525. doi: 10.1038/sj.bjp.0707003

Pryce, G., and Baker, D. 2014. Cannabis and multiple sclerosis. In *Handbook of Cannabis*, ed. R. G. Pertwee. Oxford University Press.

Qin, N., Neeper, M. P., Liu, Y., Hutchinson, T. L., Lubin, M. L., and Flores, C. M. 2008. TRPV2 is activated by cannabidiol and mediates CGRP release in cultured rat dorsal root ganglion neurons. *Journal of Neuroscience* 28 (24), 6231–6238. doi: 10.1523/JNEUROSCI.0504-08.2008

Quinn, H. R., Matsumoto, I., Callaghan, P. D., Long, L. E., Arnold, J. C., Gunasekaran, N., et al. 2008. Adolescent rats find repeated Delta (9)-THC less aversive than adult rats but display greater residual cognitive deficits and changes in hippocampal protein expression following exposure. *Neuropsychopharmacology* 33 (5), 1113–1126. doi: 10.1038/sj.npp.1301475

Rabinak, C. A., Angstadt, M., Lyons, M., Mori, S., Milad, M. R., Liberzon, I., et al. 2014. Cannabinoid modulation of prefrontal-limbic activation during fear extinction learning and recall in humans. *Neurobiology of Learning and Memory* 113, 125–134. doi: 10.1016/j.nlm.2013.09.009

Raby, W. N., Carpenter, K. M., Rothenberg, J., Brooks, A. C., Jiang, H., Sullivan, M., et al. 2009. Intermittent marijuana use is associated with improved retention in naltrexone treatment for opiate-dependence. *American Journal on Addictions* 18 (4), 301–308. doi: 10.1080/10550490902927785

Radhakrishnan, R., Wilkinson, S. T., and D'Souza, D. C. 2014. Gone to pot— A review of the association between cannabis and psychosis. *Frontiers in Psychiatry* 5, 54. doi: 10.3389/fpsyt.2014.00054

Raman, C., McAllister, S. D., Rizvi, G., Patel, S. G., Moore, D. H., and Abood, M. E. 2004. Amyotrophic lateral sclerosis: Delayed disease progression in mice by treatment with a cannabinoid. *Amyotrophic Lateral Sclerosis and Other Motor Neuron Disorders* 5 (1), 33–39. doi: 10.1080/14660820310016813

Ramer, R., and Hinz, B. 2016. Antitumorigenic targets of cannabinoids—current status and implications. *Expert Opinion on Therapeutic Targets* May 11, 1–17. doi: 10.1080/14728222.2016.1177512

Ramírez, B. G., Blazquez, C., Gomez del Pulgar, T., Guzman, M., and de Ceballos, M. L. 2005. Prevention of Alzheimer's disease pathology by cannabinoids: Neuroprotection mediated by blockade of microglial activation. *Journal of Neuroscience* 25 (8), 1904–1913. doi: 10.1523/JNEUROSCI.4540-04.2005

Ray, A. P., Griggs, L., and Darmani, N. A. 2009. Delta 9-tetrahydrocannabinol suppresses vomiting behavior and Fos expression in both acute and delayed phases of cisplatin-induced emesis in the least shrew. *Behavioural Brain Research* 196 (1), 30–36. doi: 10.1016/j.bbr.2008.07.028

Realini, N., Vigano, D., Guidali, C., Zamberletti, E., Rubino, T., and Parolaro, D. 2011. Chronic URB597 treatment at adulthood reverted most depressive-like symptoms induced by adolescent exposure to THC in female rats. *Neuropharmacology* 60 (2–3), 235–243. doi: 10.1016/j.neuropharm.2010.09.003

Reidel, G., Fadda, P., McKillop-Smith, S., Pertwee, R. G., Platt, B., and Robinson, L. 2009. Synthetic and plant-derived cannabinoid receptor antagonists show hypophagic properties in fasted and non-fasted mice. *British Journal of Pharmacology* 156, 1154–1166.

Ren, Y., Whittard, J., Higuera-Matas, A., Morris, C. V., and Hurd, Y. L. 2009. Cannabidiol, a nonpsychotropic component of cannabis, inhibits cue-induced heroin seeking and normalizes discrete mesolimbic neuronal disturbances. *Journal of Neuroscience* 29 (47), 14764–14769. doi: 10.1523/JNEUROSCI.4291-09.2009

Resstel, L. B., Joca, S. R., Moreira, F. A., Correa, F. M., and Guimaraes, F. S. 2006. Effects of cannabidiol and diazepam on behavioral and cardiovascular responses induced by contextual conditioned fear in rats. *Behavioural Brain Research* 172 (2), 294–298. doi: 10.1016/j.bbr.2006.05.016

Rey, J. M., Sawyer, M. G., Raphael, B., Patton, G. C., and Lynskey, M. 2002. Mental health of teenagers who use cannabis. Results of an Australian survey. *British Journal of Psychiatry* 180, 216–221.

Reynolds, J. R. 1868. Therapetical uses and toxic effects of Cannabis indica. *Lancet* 1, 637–638.

Riegel, A. C., and Lupica, C. R. 2004. Independent presynaptic and postsynaptic mechanisms regulate endocannabinoid signaling at multiple synapses in the ventral tegmental area. *Journal of Neuroscience* 24 (49), 11070–11078. doi: 10.1523/JNEUROSCI.3695-04.2004

Robson, P. 2011. Abuse potential and psychoactive effects of delta-9-tetrahydrocannabinol and cannabidiol oromucosal spray (Sativex), a new cannabinoid medicine. *Expert Opinion on Drug Safety* 10 (5), 675–685. doi: 10.1517/14740338.2011.575778

Roca-Pallin, J. M., Lopez-Pelayo, H., Sugranyes, G., and Balcells-Olivero, M. M. 2013. Cannabinoid hyperemesis syndrome. *CNS Neuroscience & Therapeutics* 19 (12), 994–995. doi: 10.1111/cns.12207

Rock, E. M., Bolognini, D., Limebeer, C. L., Cascio, M. G., Anavi-Goffer, S., Fletcher, P. J., et al. 2012. Cannabidiol, a non-psychotropic component of cannabis, attenuates vomiting and nausea-like behaviour via indirect agonism of 5-HT1A somatodendritic autoreceptors in the dorsal raphe nucleus. *British Journal of Pharmacology* 165 (8), 2620–2634. doi: 10.1111/j.1476-5381 .2011.01621.x

Rock, E. M., Goodwin, J. M., Limebeer, C. L., Breuer, A., Pertwee, R. G., Mechoulam, R., et al. 2011. Interaction between non-psychotropic cannabinoids in marihuana: Effect of cannabigerol (CBG) on the anti-nausea or anti-emetic effects of cannabidiol (CBD) in rats and shrews. *Psychopharmacology* 215 (3), 505–512. doi: 10.1007/s00213-010-2157-4

Rock, E. M., Kopstick, R., Limebeer, C., and Parker, L. A. 2013. Tetrahydrocannabinolic acid reduces nausea-induced conditioned gaping in rats and vomiting in Suncus murinus. *British Journal of Pharmacology* 170, 641–648.

Rock, E. M., Limebeer, C. L., Mechoulam, R., Piomelli, D., and Parker, L. A. 2008. The effect of cannabidiol and URB597 on conditioned gaping (a model of nausea) elicited by a lithium-paired context in the rat. *Psychopharmacology* 196 (3), 389–395. doi: 10.1007/s00213-007-0970-1

Rock, E. M., Limebeer, C. L., and Parker, L. A. 2014. Anticipatory nausea in animal models: A review of potential novel therapeutic treatments. *Experimental Brain Research* 232 (8), 2511–2534. doi: 10.1007/s00221-014-3942-9

Rock, E. M., Limebeer, C. L., and Parker, L. A. 2015. Effect of combined doses of Delta (9)-tetrahydrocannabinol (THC) and cannabidiolic acid (CBDA) on acute and anticipatory nausea using rat (Sprague- Dawley) models of conditioned gaping. *Psychopharmacology* 232 (24), 4445–4454. doi: 10.1007/ s00213-015-4080-.

Rock, E. M., Limebeer, C. L., Sticht, M. A., and Parker, L. A. 2014. Regulation of vomiting and nausea by phytocannabinoids. In*Handbook of Cannabis*, ed. R. G. Pertwee. Oxford University Press.

Rock, E. M., Limebeer, C. L., Ward, J. M., Cohen, A., Grove, K., Niphakis, M. J., Cravatt, B. F., and Parker, L. A. 2015. Interference with acute nausea and anticipatory nausea in rats by fatty acid amide hydrolase (FAAH) inhibition through a PPARα and CB1 receptor mechanism, respectively: a double dissociation. *Psychopharmacology* 232 (20), 3841–3848.

Rock, E. M., and Parker, L. A. 2013. Effect of low doses of cannabidiolic acid and ondansetron on LiCl-induced conditioned gaping (a model of nausea-induced behaviour) in rats. *British Journal of Pharmacology* 169 (3), 685–692.

Rock, E. M., Sticht, M. A., Duncan, M., Stott, C., and Parker, L. A. 2013. Evaluation of the potential of the phytocannabinoids, cannabidivarin (CBDV) and Delta-tetrahydrocannabivarin (THCV), to produce CB receptor inverse agonism symptoms of nausea in rats. *British Journal of Pharmacology* 170 (3), 671–678.

Rock, E. M., Sticht, M. A., and Parker, L. A. 2014. Effect of phytocannabinoids on nausea and vomiting. In *Handbook of Cannabis*, ed. R. G. Pertwee. Oxford University Presss.

Rog, D. J., Nurmikko, T. J., Friede, T., and Young, C. A. 2005. Randomized, controlled trial of cannabis-based medicine in central pain in multiple sclerosis. *Neurology* 65 (6), 812–819. doi: 10.1212/01.wnl.0000176753.45410.8b

Rog, D. J., Nurmikko, T. J., and Young, C. A. 2007. Oromucosal delta9-tetrahydrocannabinol/cannabidiol for neuropathic pain associated with multiple sclerosis: An uncontrolled, open-label, 2-year extension trial. *Clinical Therapeutics* 29 (9), 2068–2079. doi: 10.1016/j.clinthera.2007.09.013

Rogeberg, O., and Elvik, R. 2016. The effects of cannabis intoxication on motor vehicle collision revisited and revised. *Addiction* 111 (8), 1348–1359.

Roitman, P., Mechoulam, R., Cooper-Kazaz, R., and Shalev, A. 2014. Preliminary, open-label, pilot study of add-on oral Delta9-tetrahydrocannabinol in chronic post-traumatic stress disorder. *Clinical Drug Investigation* 34 (8), 587–591. doi: 10.1007/s40261-014-0212-3

Romigi, A., Bari, M., Placidi, F., Marciani, M. G., Malaponti, M., Torelli, F., et al. 2010. Cerebrospinal fluid levels of the endocannabinoid anandamide are reduced in patients with untreated newly diagnosed temporal lobe epilepsy. *Epilepsia* 51 (5), 768–772. doi: 10.1111/j.1528-1167.2009.02334.x

Rosenblum, A., Marsch, L. A., Joseph, H., and Portenoy, R. K. 2008. Opioids and the treatment of chronic pain: Controversies, current status, and future directions. *Experimental and Clinical Psychopharmacology* 16 (5), 405–416. doi: 10.1037/a0013628

Rubino, T., Massi, P., Vigano, D., Fuzio, D., and Parolaro, D. 2000. Long-term treatment with SR141716A, the CB1 receptor antagonist, influences morphine withdrawal syndrome. *Life Sciences* 66 (22), 2213–2219.

Rubino, T., and Parolaro, D. 2008. Long lasting consequences of cannabis exposure in adolescence. Molecular and Cellular Endocrinology 286 (1–2 Suppl 1), S108–S113. doi: 10.1016/j.mce.2008.02.003

Ruehle, S., Remmers, F., Romo-Parra, H., Massa, F., Wickert, M., Wortge, S., et al. 2013. Cannabinoid CB1 receptor in dorsal telencephalic glutamatergic neurons: Distinctive sufficiency for hippocampus-dependent and amygdala-dependent synaptic and behavioral functions. *Journal of Neuroscience* 33 (25), 10264–10277. doi: 10.1523/JNEUROSCI.4171-12.2013

Ruiz-Veguilla, M., Barrigon, M. L., Hernandez, L., Rubio, J. L., Gurpegui, M., Sarramea, F., et al. 2013. Dose-response effect between cannabis use and psychosis liability in a non-clinical population: Evidence from a snowball sample. *Journal of Psychiatric Research* 47 (8), 1036–1043. doi: 10.1016/j.jpsychires.2013.03.003

Russo, E. B. 2014. The pharmacological history of cannabis. In *Handbook of Cannabis*, ed. R. G. Pertwee. Oxford University Press.

Ryberg, E., Larsson, N., Sjogren, S., Hjorth, S., Hermansson, N. O., Leonova, J., et al. 2007. The orphan receptor GPR55 is a novel cannabinoid receptor. *British Journal of Pharmacology* 152 (7), 1092–1101. doi: 10.1038/sj.bjp.0707460

Sachse-Seeboth, C., Pfeil, J., Sehrt, D., Meineke, I., Tzvetkov, M., Bruns, E., et al. 2009. Interindividual variation in the pharmacokinetics of Delta9-tetrahydrocannabinol as related to genetic polymorphisms in CYP2C9. *Clinical Pharmacology and Therapeutics* 85 (3), 273–276. doi: 10.1038/clpt.2008.213

Sagredo, O., Ramos, J. A., Decio, A., Mechoulam, R., and Fernandez-Ruiz, J. 2007) Cannabidiol reduced the striatal atrophy caused 3-nitropropionic acid in vivo by mechanisms independent of the activation of cannabinoid, vanilloid TRPV1 and adenosine A2A receptors. *European Journal of Neuroscience* 26 (4), 843–851. doi: 10.1111/j.1460-9568.2007.05717.x

Sanchez, M. G., Ruiz-Llorente, L., Sanchez, A. M., and Diaz-Laviada, I. 2003. Activation of phosphoinositide 3-kinase/PKB pathway by CB (1) and CB (2) cannabinoid receptors expressed in prostate PC-3 cells. Involvement in Raf-1 stimulation and NGF induction. *Cellular Signalling* 15 (9), 851–859.

Sasso, O., Bertorelli, R., Bandiera, T., Scarpelli, R., Colombano, G., Armirotti, A., et al. 2012. Peripheral FAAH inhibition causes profound antinociception and protects against indomethacin-induced gastric lesions. *Pharmacological Research*, 65 (5), 553–563. doi: 10.1016/j.phrs.2012.02.012

Savage, S. R., Romero-Sandoval, A., Schatman, M., Wallace, M., Fanciullo, G., McCarberg, B., et al. 2016. Cannabis in pain treatment: Clinical and research considerations. *Journal of Pain* 17 (6), 654–668.

Sawler, J., Stout, J. M., Gardner, K. M., Hudson, D., Vidmar, J., Butler, L., et al. 2015. The genetic structure of marijuana and hemp. *PLoS One* 10 (8), e0133292. doi: 10.1371/journal.pone.0133292

Sawzdargo, M., Nguyen, T., Lee, D. K., Lynch, K. R., Cheng, R., Heng, H. H., et al. 1999. Identification and cloning of three novel human G protein-coupled receptor genes GPR52, PsiGPR53 and GPR55: GPR55 is extensively expressed in human brain. *Brain Research. Molecular Brain Research* 64 (2), 193–198.

Scavone, J. L., Sterling, R. C., and Van Bockstaele, E. J. 2013. Cannabinoid and opioid interactions: Implications for opiate dependence and withdrawal. *Neuroscience* 248, 637–654. doi: 10.1016/j.neuroscience.2013.04.034

Scheen, A. J., Finer, N., Hollander, P., Jensen, M. D., Van Gaal, L. F., et al. 2006. Efficacy and tolerability of rimonabant in overweight or obese patients with type 2 diabetes: A randomised controlled study. *Lancet* 368 (9548), 1660–1672. doi: 10.1016/S0140-6736(06)69571-8

Schelling, G., Hauer, D., Azad, S. C., Schmoelz, M., Chouker, A., Schmidt, M., et al. 2006. Effects of general anesthesia on anandamide blood levels in humans. *Anesthesiology* 104 (2), 273–277.

Scherma, M., Panlilio, L. V., Fadda, P., Fattore, L., Gamaleddin, I., Le Foll, B., et al. 2008. Inhibition of anandamide hydrolysis by cyclohexyl carbamic acid 3′-carbamoyl-3-yl ester (URB597) reverses abuse-related behavioral and neurochemical effects of nicotine in rats. *Journal of Pharmacology and Experimental Therapeutics* 327 (2), 482–490. doi: 10.1124/jpet.108.142224

Schierenbeck, T., Riemann, D., Berger, M., and Hornyak, M. 2008. Effect of illicit recreational drugs upon sleep: Cocaine, ecstasy and marijuana. *Sleep Medicine Reviews* 12 (5), 381–389. doi: 10.1016/j.smrv.2007.12.004

Schimmelmann, B. G., Conus, P., Cotton, S., Kupferschmid, S., McGorry, P. D., and Lambert, M. 2012. Prevalence and impact of cannabis use disorders in adolescents with early onset first episode psychosis. *European Psychiatry* 27 (6), 463–469. doi: 10.1016/j.eurpsy.2011.03.001

Schlosburg, J. E., Blankman, J. L., Long, J. Z., Nomura, D. K., Pan, B., Kinsey, S. G., et al. 2010. Chronic monoacylglycerol lipase blockade causes functional antagonism of the endocannabinoid system. *Nature Neuroscience* 13 (9), 1113–1119. doi: 10.1038/nn.2616

Schneider, M., and Koch, M. 2003. Chronic pubertal, but not adult chronic cannabinoid treatment impairs sensorimotor gating, recognition memory, and the performance in a progressive ratio task in adult rats. *Neuropsychopharmacology* 28 (10), 1760–1769. doi: 10.1038/sj.npp.1300225

Schoedel, K. A., Chen, N., Hilliard, A., White, L., Stott, C., Russo, E., et al. 2011. A randomized, double-blind, placebo-controlled, crossover study to evaluate the subjective abuse potential and cognitive effects of nabiximols oromucosal spray in subjects with a history of recreational cannabis use. *Human Psychopharmacology* 26 (3), 224–236. doi: 10.1002/hup.1196

Schreiner, A. M., and Dunn, M. E. 2012. Residual effects of cannabis use on neurocognitive performance after prolonged abstinence: A meta-analysis. *Experimental and Clinical Psychopharmacology* 20 (5), 420–429. doi: 10.1037/a0029117

Schubart, C. D., Sommer, I. E., van Gastel, W. A., Goetgebuer, R. L., Kahn, R. S., and Boks, M. P. 2011. Cannabis with high cannabidiol content is associated with fewer psychotic experiences. *Schizophrenia Research* 130 (1–3), 216–221. doi: 10.1016/j.schres.2011.04.017

Schubart, C. D., van Gastel, W. A., Breetvelt, E. J., Beetz, S. L., Ophoff, R. A., Sommer, I. E., et al. 2011. Cannabis use at a young age is associated with psychotic experiences. *Psychological Medicine* 41 (6), 1301–1310. doi: 10.1017/S003329171000187X

Sciolino, N. R., Zhou, W., and Hohmann, A. G. 2011. Enhancement of endocannabinoid signaling with JZL184, an inhibitor of the 2-arachidonoylglycerol hydrolyzing enzyme monoacylglycerol lipase, produces anxiolytic effects under conditions of high environmental aversiveness in rats. *Pharmacological Research* 64 (3), 226–234. doi: 10.1016/j.phrs.2011.04.010

Scopinho, A. A., Guimaraes, F. S., Correa, F. M., and Resstel, L. B. 2011. Cannabidiol inhibits the hyperphagia induced by cannabinoid-1 or serotonin-1A receptor agonists. *Pharmacology, Biochemistry, and Behavior* 98 (2), 268–272. doi: 10.1016/j.pbb.2011.01.007

Serpell, M., Ratcliffe, S., Hovorka, J., Schofield, M., Taylor, L., Lauder, H., et al. 2014. A double-blind, randomized, placebo-controlled, parallel group study of THC/CBD spray in peripheral neuropathic pain treatment. *European Journal of Pain* 18 (7), 999–1012. doi: 10.1002/j.1532-2149.2013.00445.x

Serrano, A., and Parsons, L. H. 2011. Endocannabinoid influence in drug reinforcement, dependence and addiction-related behaviors. *Pharmacology & Therapeutics* 132 (3), 215–241. doi: 10.1016/j.pharmthera.2011.06.005

Sharkey, K. A., Darmani, N. A., and Parker, L. A. 2014. Regulation of nausea and vomiting by cannabinoids and the endocannabinoid system. *European Journal of Pharmacology* 722, 134–146.

Shelef, A., Barak, Y., Berger, U., Paleacu, D., Tadger, S., Plopsky, I., et al. 2016. Safety and efficacy of medical cannabis oil for behavioral and psychological symptoms of dementia: An open label, add-on, pilot study. *Journal of Alzheimer's Disease* 51 (1), 15–19. doi: 10.3233/JAD-150915

Shen, M., and Thayer, S. A. 1998. Cannabinoid receptor agonists protect cultured rat hippocampal neurons from excitotoxicity. *Molecular Pharmacology* 54 (3), 459–462.

Siegel, S. (in press). The heroin overdose mystery. *Current Directions in Psychological Science.*

Simonetto, D. A., Oxentenko, A. S., Herman, M. L., and Szostek, J. H. 2012. Cannabinoid hyperemesis: A case series of 98 patients. *Mayo Clinic Proceedings* 87 (2), 114–119. doi: 10.1016/j.mayocp.2011.10.005

Sim-Selley, L. J. 2003. Regulation of cannabinoid CB_1 receptors in the central nervous system by chronic cannabinoids. *Critical Reviews in Neurobiology* 15 (2), 91–119.

Sink, K. S., McLaughlin, P. J., Wood, J. T., Brown, C., Fan, P., Vemuri, V. K., et al. 2008. The novel cannabinoid CB_1 receptor neutral antagonist AM4113 suppresses food intake and food-reinforced behavior but does not induce signs of nausea in rats. *Neuropsychopharmacology* 33 (4), 946–955. doi: 10.1038/sj.npp.1301476

Sink, K. S., Segovia, K. N., Collins, L. E., Markus, E. J., Vemuri, V. K., Makriyannis, A., et al. 2010. The CB1 inverse agonist AM251, but not the CB1 antagonist AM4113, enhances retention of contextual fear conditioning in rats. *Pharmacology, Biochemistry, and Behavior* 95 (4), 479–484. doi: 10.1016/j.pbb.2010.03.011

Sink, K. S., Segovia, K. N., Sink, J., Randall, P. A., Collins, L. E., Correa, M., et al. 2010. Potential anxiogenic effects of cannabinoid CB$_1$ receptor antagonists/inverse agonists in rats: Comparisons between AM4113, AM251, and the benzodiazepine inverse agonist FG-7142. *European Neuropsychopharmacology* 20 (2), 112–122. doi: 10.1016/j.euroneuro.2009.11.00.

Skosnik, P. D., Cortes-Briones, J. A., and Hajos, M. 2016. It's all in the rhythm: The role of cannabinoids in neural oscillations and psychosis. *Biological Psychiatry* 79 (7), 568–577. doi: 10.1016/j.biopsych.2015.12.011

Sofia, R. D., and Knobloch, L. C. 1976. Comparative effects of various naturally occurring cannabinoids on food, sucrose and water consumption by rats. *Pharmacology, Biochemistry, and Behavior* 4 (5), 591–599.

Sofia, R. D., Vassar, H. B., and Knobloch, L. C. 1975. Comparative analgesic activity of various naturally occurring cannabinoids in mice and rats. *Psychopharmacology* 40 (4), 285–295.

Solowij, N., and Battisti, R. 2008. The chronic effects of cannabis on memory in humans: A review. *Current Drug Abuse Reviews* 1 (1), 81–98.

Soltesz, I., Alger, B. E., Kano, M., Lee, S. H., Lovinger, D. M., Ohno-Shosaku, T., et al. 2015. Weeding out bad waves: Towards selective cannabinoid circuit control in epilepsy. *Nature Reviews. Neuroscience* 16 (5), 264–277. doi: 10.1038/nrn3937

Spoto, B., Fezza, F., Parlongo, G., Battista, N., Sgro, E., Gasperi, V., et al. 2006. Human adipose tissue binds and metabolizes the endocannabinoids anandamide and 2-arachidonoylglycerol. *Biochimie* 88 (12), 1889–1897. doi: 10.1016/j.biochi.2006.07.019

Stefanis, N. C., Dragovic, M., Power, B. D., Jablensky, A., Castle, D., and Morgan, V. A. 2013. Age at initiation of cannabis use predicts age at onset of psychosis: The 7- to 8-year trend. *Schizophrenia Bulletin* 39 (2), 251–254. doi: 10.1093/schbul/sbs188

Stella, N. 2004. Cannabinoid signaling in glial cells. *Glia* 48 (4), 267–277. doi: 10.1002/glia.20084

Sticht, M. A., Limebeer, C. L., Rafla, B. R., Abdullah, R. A., Poklis, J. L., Ho, W., et al. 2015. Endocannabinoid regulation of nausea is mediated by 2-arachidonoylglycerol (2-AG) in the rat visceral insular cortex. *Neuropharmacology* 102, 92–102.

Sticht, M. A., Limebeer, C. L., Rafla, B. R., and Parker, L. A. 2015. Intra-visceral insular cortex 2-arachidonoylglycerol, but not N-arachidonoylethanolamide, suppresses acute nausea-induced conditioned gaping in rats. *Neuroscience* 286, 338–344. doi: 10.1016/j.neuroscience.2014.11.058

Sticht, M. A., Rock, E. M., Limebeer, C. L., & Parker, L. A. (2015). Endocannabinoid mechanisms influencing nausea. *International Review of Neurobiology*, 125, 127–162.

Stokes, P. R., Mehta, M. A., Curran, H. V., Breen, G., and Grasby, P. M. 2009. Can recreational doses of THC produce significant dopamine release in the human striatum? *NeuroImage* 48 (1), 186–190. doi: 10.1016/j.neuroimage.2009.06.029

Stover, J. F., Pleines, U. E., Morganti-Kossmann, M. C., Kossmann, T., Lowitzsch, K., and Kempski, O. S. 1997. Neurotransmitters in cerebrospinal fluid reflect pathological activity. *European Journal of Clinical Investigation* 27 (12), 1038–1043.

Substance Abuse and Mental Health Services Administration. 2013. Results from the 2013 national surevey on drug use and health: Summary of national findings. http://store.samhsa.gov/product/Results-from-the-2013-National-Survey-on-Drug-Use-and-Health-Summary-of-National-Findings/SMA144863.

Sugiura, T., Kondo, S., Sukagawa, A., Nakane, S., Shinoda, A., Itoh, K., et al. 1995. 2-Arachidonoylglycerol: A possible endogenous cannabinoid receptor ligand in brain. *Biochemical and Biophysical Research Communications* 215 (1), 89–97.

Sulcova, E., Mechoulam, R., and Fride, E. 1998. Biphasic effects of anandamide. *Pharmacology, Biochemistry, and Behavior* 59 (2), 347–352.

Sulkowski, G., Dabrowska-Bouta, B., Kwiatkowska-Patzer, B., and Struzynska, L. 2009. Alterations in glutamate transport and group I metabotropic glutamate receptors in the rat brain during acute phase of experimental autoimmune encephalomyelitis. *Folia Neuropathologica* 47 (4), 329–337.

Sullivan, E. V. 2007. Alcohol and drug dependence: Brain mechanisms and behavioral impact. *Neuropsychology Review* 17 (3), 235–238. doi: 10.1007/s11065-007-9039-5

Sullivan, S. 2010. Cannabinoid hyperemesis. *Canadian Journal of Gastroenterology* 24 (5), 284–285.

Sumislawski, J. J., Ramikie, T. S., and Patel, S. 2011. Reversible gating of endocannabinoid plasticity in the amygdala by chronic stress: A potential role for monoacylglycerol lipase inhibition in the prevention of stress-induced behavioral adaptation. *Neuropsychopharmacology* 36 (13), 2750–2761. doi: 10.1038/npp.2011.166

Sutton, I. R., and Daeninck, P. 2006. Cannabinoids in the management of intractable chemotherapy-induced nausea and vomiting and cancer-related pain. *Journal of Supportive Oncology* 4 (10), 531–535.

Suzuki, A., Mukawa, T., Tsukagoshi, A., Frankland, P. W., and Kida, S. 2008. Activation of LVGCCs and CB_1 receptors required for destabilization of reactivated contextual fear memories. *Learning & Memory* 15 (6), 426–433. doi: 10.1101/lm.888808

Sylantyev, S., Jensen, T. P., Ross, R. A., & Rusakov, D. A. (2013). Cannabinoid- and lysophosphatidylinositol-sensitive receptor GPR55 boost neurotransmitter release at central synapses. *Proceedings of the National Academy of Sciences of the United States of America*, 110(13), 5193–5198.

Tam, J., Cinar, R., Liu, J., Godlewski, G., Wesley, D., Jourdan, T., et al. 2012. Peripheral cannabinoid-1 receptor inverse agonism reduces obesity by reversing leptin resistance. *Cell Metabolism* 16 (2), 167–179. doi: 10.1016/j.cmet .2012.07.002

Tam, J., Vemuri, V. K., Liu, J., Batkai, S., Mukhopadhyay, B., Godlewski, G., et al. 2010. Peripheral CB1 cannabinoid receptor blockade improves cardiometabolic risk in mouse models of obesity. *Journal of Clinical Investigation* 120 (8), 2953–2966. doi: 10.1172/JCI42551

Tan, H., Lauzon, N. M., Bishop, S. F., Chi, N., Bechard, M., and Laviolette, S. R. 2011. Cannabinoid transmission in the basolateral amygdala modulates fear memory formation via functional inputs to the prelimbic cortex. *Journal of Neuroscience* 31 (14), 5300–5312. doi: 10.1523/JNEUROSCI.4718-10.2011

Ternianov, A., Perez-Ortiz, J. M., Solesio, M. E., Garcia-Gutierrez, M. S., Ortega-Alvaro, A., Navarrete, F., et al. (2012). Overexpression of CB2 cannabinoid receptors results in neuroprotection against behavioral and neurochemical alterations induced by intracaudate administration of 6-hydroxydopamine. *Neurobiology of Aging*, 33(2), 421.e1–421.e16. doi:10.1016/j.neurobiolaging .2010.09.012.

Thomas, A., Stevenson, L. A., Wease, K. N., Price, M. R., Baillie, G., Ross, R. A., et al. 2005. Evidence that the plant cannabinoid Delta9-tetrahydrocannabivarin is a cannabinoid CB1 and CB2 receptor antagonist. *British Journal of Pharmacology* 146 (7), 917–926. doi: 10.1038/sj.bjp.0706414

Thomas, B. F., Wiley, J. L., Pollard, G. T., and Grabenauer, M. 2014. Cannabinoid Designer Drugs: Effects and Forensics. In *Handbook of Cannabis*, ed. R. G. Pertwee. Oxford University Press.

Todd, A. R. 1946. Hashish. *Experientia* 2, 55–60.

Tourino, C., Oveisi, F., Lockney, J., Piomelli, D., and Maldonado, R. 2010. FAAH deficiency promotes energy storage and enhances the motivation for food. *International Journal of Obesity* 34 (3), 557–568. doi: 10.1038/ijo.2009.262

Tramer, M. R., Carroll, D., Campbell, F. A., Reynolds, D. J., Moore, R. A., and McQuay, H. J. 2001. Cannabinoids for control of chemotherapy induced nausea and vomiting: Quantitative systematic review. *BMJ* 323 (7303), 16–21.

Travers, J. B., and Norgren, R. 1986. Electromyographic analysis of the ingestion and rejection of sapid stimuli in the rat. *Behavioral Neuroscience* 100 (4), 544–555.

Trigo, J. M., and Le Foll, B. 2015. Inhibition of monoacylglycerol lipase (MAGL) enhances cue-induced reinstatement of nicotine-seeking behavior in mice. *Psychopharmacology* 233 (10), 1815–1822. doi: 10.1007/s00213-015-4117-5

Tuerke, K. J., Limebeer, C. L., Fletcher, P. J., and Parker, L. A. 2012. Double dissociation between regulation of conditioned disgust and taste avoidance by serotonin availability at the 5-HT (3) receptor in the posterior and anterior insular cortex. *Journal of Neuroscience* 32 (40), 13709–13717.

Turkanis, S. A., Smiley, K. A., Borys, H. K., Olsen, D. M., and Karler, R. 1979. An electrophysiological analysis of the anticonvulsant action of cannabidiol on limbic seizures in conscious rats. *Epilepsia* 20 (4), 351–363.

Valjent, E., and Maldonando, R. 2000. A behavioral model to reveal pace preference to delta-9-tetrahydrocannabinol in mice. *Psychopharmacology* 147 (4), 436–438.

Vallee, M., Vitiello, S., Bellocchio, L., Hebert-Chatelain, E., Monlezun, S., Martin-Garcia, E., et al. 2014. Pregnenolone can protect the brain from cannabis intoxication. *Science* 343 (6166), 94–98. doi: 10.1126/science.1243985

van Anders, S. M., Chernick, A. B., Chernick, B. A., Hampson, E., and Fisher, W. A. 2005. Preliminary clinical experience with androgen administration for pre- and postmenopausal women with hypoactive sexual desire. *Journal of Sex & Marital Therapy* 31 (3), 173–185. doi: 10.1080/00926230590513384

Van Belle, S., Lichinitser, M. R., Navari, R. M., Garin, A. M., Decramer, M. L., Riviere, A., et al. 2002. Prevention of cisplatin-induced acute and delayed emesis by the selective neurokinin-1 antagonists, L-758,298 and MK-869. *Cancer* 94 (11), 3032–3041. doi: 10.1002/cncr.10516

van den Elsen, G. A., Ahmed, A. I., Verkes, R. J., Kramers, C., Feuth, T., Rosenberg, P. B., et al. 2015. Tetrahydrocannabinol for neuropsychiatric symptoms in dementia: A randomized controlled trial. *Neurology* 84 (23), 2338–2346. doi: 10.1212/WNL.0000000000001675

van der Pol, P., Liebregts, N., Brunt, T., van Amsterdam, J., de Graaf, R., Korf, D. J., et al. 2014. Cross-sectional and prospective relation of cannabis potency, dosing and smoking behaviour with cannabis dependence: An ecological study. *Addiction* 109 (7), 1101–1109. doi: 10.1111/add.12508

van der Stelt, M., and Di Marzo, V. 2005. Anandamide as an intracellular messenger regulating ion channel activity. *Prostaglandins & Other Lipid Mediators* 77 (1–4), 111–122. doi: 10.1016/j.prostaglandins.2004.09.007

Van Gaal, L. F., Rissanen, A. M., Scheen, A. J., Ziegler, O., Rossner, S., et al. 2005. Effects of the cannabinoid-1 receptor blocker rimonabant on weight reduction and cardiovascular risk factors in overweight patients: 1-year experience from the RIO-Europe study. *Lancet* 365 (9468), 1389–1397. doi: 10.1016/S0140-6736(05)66374-X

Van Sickle, M. D., Duncan, M., Kingsley, P. J., Mouihate, A., Urbani, P., Mackie, K., et al. 2005. Identification and functional characterization of brainstem cannabinoid CB$_2$ receptors. *Science* 310 (5746), 329–332.

van Os, J., and Kapur, S. 2009. Schizophrenia. *Lancet*, 374 (9690), 635–645. doi: 10.1016/S0140-6736(09)60995-8

Van Sickle, M. D., Oland, L. D., Ho, W., Hillard, C. J., Mackie, K., Davison, J. S., et al. 2001. Cannabinoids inhibit emesis through CB$_1$ receptors in the brainstem of the ferret. *Gastroenterology* 121 (4), 767–774.

van Winkel, R., van Beveren, N. J., Simons, C., Genetic Risk and Outcome of Psychosis (GROUP) Investigators, and Genetic Risk and Outcome of Psychosis (GROUP) Investigators. 2011. AKT1 moderation of cannabis-induced cognitive alterations in psychotic disorder. *Neuropsychopharmacology* 36 (12), 2529–2537. doi: 10.1038/npp.2011.141

Varvel, S. A., Hamm, R. J., Martin, B. R., and Lichtman, A. H. 2001. Differential effects of delta 9-THC on spatial reference and working memory in mice. *Psychopharmacology* 157 (2), 142–150.

Varvel, S. A., and Lichtman, A. H. 2002. Evaluation of CB₁ receptor knockout mice in the Morris water maze. *Journal of Pharmacology and Experimental Therapeutics* 301 (3), 915–924.

Varvel, S. A., Wiley, J. L., Yang, R., Bridgen, D. T., Long, K., Lichtman, A. H., et al. 2006. Interactions between THC and cannabidiol in mouse models of cannabinoid activity. *Psychopharmacology* 186 (2), 226–234. doi: 10.1007/s00213-006-0356-9

Varvel, S. A., Wise, L. E., and Lichtman, A. H. 2009. Are CB (1) receptor antagonists nootropic or cognitive impairing agents? *Drug Development Research* 70 (8), 555–565. doi: 10.1002/ddr.20334

Varvel, S. A., Wise, L. E., Niyuhire, F., Cravatt, B. F., and Lichtman, A. H. 2007. Inhibition of fatty-acid amide hydrolase accelerates acquisition and extinction rates in a spatial memory task. *Neuropsychopharmacology* 32 (5), 1032–1041. doi: 10.1038/sj.npp.1301224

Verhoeckx, K. C., Korthout, H. A., van Meeteren-Kreikamp, A. P., Ehlert, K. A., Wang, M., van der Greef, J., et al. 2006. Unheated *Cannabis sativa* extracts and its major compound THC-acid have potential immuno-modulating properties not mediated by CB₁ and CB₂ receptor coupled pathways. *International Immunopharmacology* 6 (4), 656–665. doi: 10.1016/j.intimp.2005.10.002

Vigano, D., Rubino, T., Vaccani, A., Bianchessi, S., Marmorato, P., Castiglioni, C., et al. 2005. Molecular mechanisms involved in the asymmetric interaction between cannabinoid and opioid systems. *Psychopharmacology* 182 (4), 527–536. doi: 10.1007/s00213-005-0114-4

Vos, T., Flaxman, A. D., Naghavi, M., Lozano, R., Michaud, C., Ezzati, M., et al. 2012. Years lived with disability (YLDs) for 1160 sequelae of 289 diseases and injuries 1990–2010: A systematic analysis for the Global Burden of Disease Study 2010. *Lancet* 380 (9859), 2163–2196. doi: 10.1016/S0140-6736(12)61729-2

Wachtel, S. R., ElSohly, M. A., Ross, S. A., Ambre, J., and de Wit, H. 2002. Comparison of the subjective effects of Delta (9)-tetrahydrocannabinol and marijuana in humans. *Psychopharmacology* 161 (4), 331–339. doi: 10.1007/s00213-002-1033-2

Wade, D. T., Makela, P. M., House, H., Bateman, C., and Robson, P. 2006. Long-term use of a cannabis-based medicine in the treatment of spasticity and other symptoms in multiple sclerosis. *Multiple Sclerosis* 12 (5), 639–645.

Waissengrin, B., Urban, D., Leshem, Y., Garty, M., and Wolf, I. 2015. Patterns of use of medical cannabis among Israeli cancer patients: A single institution experience. *Journal of Pain and Symptom Management* 49 (2), 223–230. doi: 10.1016/j.jpainsymman.2014.05.018

Wallace, M. J., Blair, R. E., Falenski, K. W., Martin, B. R., and DeLorenzo, R. J. 2003. The endogenous cannabinoid system regulates seizure frequency and duration in a model of temporal lobe epilepsy. *Journal of Pharmacology and Experimental Therapeutics* 307 (1), 129–137. doi: 10.1124/jpet.103.051920

Ward, S. J., McAllister, S. D., Kawamura, R., Murase, R., Neelakantan, H., and Walker, E. A. 2014. Cannabidiol inhibits paclitaxel-induced neuropathic pain through 5-HT (1A) receptors without diminishing nervous system function or chemotherapy efficacy. *British Journal of Pharmacology* 171 (3), 636–645. doi: 10.1111/bph.12439

Ware, M. A., Adams, H., and Guy, G. W. 2005. The medicinal use of cannabis in the UK: Results of a nationwide survey. *International Journal of Clinical Practice* 59 (3), 291–295. doi: 10.1111/j.1742-1241.2004.00271.x

Ware, M. A., Fitzcharles, M. A., Joseph, L., and Shir, Y. 2010. The effects of nabilone on sleep in fibromyalgia: Results of a randomized controlled trial. *Anesthesia and Analgesia* 110 (2), 604–610. doi: 10.1213/ANE.0b013e3181c76f70

Ware, M. A., and Tawfik, V. L. 2005. Safety issues concerning the medical use of cannabis and cannabinoids. *Pain Res Manag, 10 Suppl A*, 31A–37A.

Ware, M. A., Wang, T., Shapiro, S., Collet, J. P., et al. 2015. Cannabis for the Management of Pain: Assessment of Safety Study (COMPASS). *Journal of Pain* 16 (12), 1233–1242. doi: 10.1016/j.jpain.2015.07.014

Weiland, B. J., Thayer, R. E., Depue, B. E., Sabbineni, A., Bryan, A. D., and Hutchison, K. E. 2015. Daily marijuana use is not associated with brain morphometric measures in adolescents or adults. *Journal of Neuroscience* 35 (4), 1505–1512. doi: 10.1523/JNEUROSCI.2946-14.2015

Wendt, H., Soerensen, J., Wotjak, C. T., and Potschka, H. 2011. Targeting the endocannabinoid system in the amygdala kindling model of temporal lobe epilepsy in mice. *Epilepsia* 52 (7), e62–e65. doi: 10.1111/j.1528-1167.2011.03079.x

Whiting, P. F., Wolff, R. F., Deshpande, S., Di Nisio, M., Duffy, S., Hernandez, A. V., et al. 2015. Cannabinoids for medical use: A systematic review and meta-analysis. *Journal of the American Medical Association* 313 (24), 2456–2473. doi: 10.1001/jama.2015.6358

Wiley, J. L., Burston, J. J., Leggett, D. C., Alekseeva, O. O., Razdan, R. K., Mahadevan, A., et al. 2005. CB_1 cannabinoid receptor-mediated modulation of food intake in mice. *British Journal of Pharmacology* 145 (3), 293–300. doi: 10.1038/sj.bjp.0706157

Wilkinson, S. T., Radhakrishnan, R., and D'Souza, D. C. 2014. Impact of cannabis use on the development of psychotic disorders. *Current Addiction Reports* 1 (2), 115–128. doi: 10.1007/s40429-014-0018-7

Williams, C. M., Jones, N. A., and Whalley, B. J. 2014. Cannabis and epilepsy. In *Handbook of Cannabis*, ed. R. G. Pertwee. Oxford University Press.

Wills, K. L., Vemuri, K., Kalmar, A., Lee, A., Limebeer, C. L., Makriyannis, A., and Parker, L. A. 2014. CB$_1$ antagonists interfere with the establishment of a one trial naloxone-precipitated morphine withdrawal induced place aversion. *Psychopharmacology* 231, 4291–4300.

Wills, K. L., Limebeer, C. L., Rock, E. M., Niphakis, M. J., Cravatt, B. F., and Parker, L. A. 2016. Double dissociation of monoacylglycerol lipase inhibition and CB1 antagonism in the basolateral amygdala, central amygdala and the interoceptive insular cortex on the affective properties of acute naloxone-precipitated morphine withdrawal in rats. *Neuropsychopharmacology* 41, 1865–1873.

Wills, K. L., and Parker, L. A. 2016. Effect of pharmacological modulation of the endocannabinoid system on opiate withdrawal: A review of the preclinical animal literature. *Frontiers in Pharmacology* 28 (7), 187. doi: 10.3389/fphar.2016.00187.

Wilsey, B., Marcotte, T., Deutsch, R., Gouaux, B., Sakai, S., and Donaghe, H. 2013. Low-dose vaporized cannabis significantly improves neuropathic pain. *Journal of Pain* 14 (2), 136–148. doi: 10.1016/j.jpain.2012.10.009

Wilsey, B., Marcotte, T., Tsodikov, A., Millman, J., Bentley, H., Gouaux, B., et al. 2008. A randomized, placebo-controlled, crossover trial of cannabis cigarettes in neuropathic pain. *Journal of Pain* 9 (6), 506–521. doi: 10.1016/j.jpain.2007.12.010

Wilson, R. I., and Nichol, R. A. 2001. Endogenous cannabinoids mediate retrograde signaling at hippocampal synapses. *Nature*, 410, 588–592.

Wise, L. E., Long, K. A., Abdullah, R. A., Long, J. Z., Cravatt, B. F., and Lichtman, A. H. 2012. Dual fatty acid amide hydrolase and monoacylglycerol lipase blockade produces THC-like Morris water maze deficits in mice. *ACS Chemical Neuroscience* 3 (5), 369–378. doi: 10.1021/cn200130s

Wise, R. A. 2004. Dopamine, learning and motivation. *Nature Reviews Neuroscience* 5 (6), 483–494. doi: 10.1038/nrn1406

Wood, T. B., Spivey, W. T. N., and Easterfield, T. H. 1896. Charas, the resin of Indian hemp. *Journal of the Chemical Society* 69, 539–546.

Woodhams, S. G., Sagar, D. R., & Chapman, V. (2015). The role of the endocannabinoid system in pain. *Handbook of Experimental Pharmacology*, 227, 119–143.

Xi, Z. X., Peng, X. Q., Li, X., Song, R., Zhang, H. Y., Liu, Q. R., et al. 2011. Brain cannabinoid CB (2) receptors modulate cocaine's actions in mice. *Nature Neuroscience* 14 (9), 1160–1166. doi: 10.1038/nn.2874

Xiong, W., Cui, T., Cheng, K., Yang, F., Chen, S. R., Willenbring, D., et al. 2012. Cannabinoids suppress inflammatory and neuropathic pain by targeting alpha3 glycine receptors. *Journal of Experimental Medicine* 209 (6), 1121–1134. doi: 10.1084/jem.20120242

Yarnell, S. 2015. The use of medicinal marijuana for posttraumatic stress disorder: A review of the current literature. *Primary Care Companion for CNS Disorders* 17 (3). doi: 10.4088/PCC.15r01786

Yasui, Y., Breder, C. D., Saper, C. B., and Cechetto, D. F. 1991. Autonomic responses and efferent pathways from the insular cortex in the rat. *Journal of Comparative Neurology* 303 (3), 355–374.

Yucel, M., Bora, E., Lubman, D. I., Solowij, N., Brewer, W. J., Cotton, S. M., et al. 2012. The impact of cannabis use on cognitive functioning in patients with schizophrenia: A meta-analysis of existing findings and new data in a first-episode sample. *Schizophrenia Bulletin* 38 (2), 316–330. doi: 10.1093/schbul/sbq079

Zajicek, J., Ball, S., Wright, D., Vickery, J., Nunn, A., Miller, D., et al. 2013. Effect of dronabinol on progression in progressive multiple sclerosis (CUPID): A randomised, placebo-controlled trial. *Lancet Neurology* 12 (9), 857–865. doi: 10.1016/S1474-4422(13)70159-5

Zammit, S., Allebeck, P., Andreasson, S., Lundberg, I., and Lewis, G. 2002. Self reported cannabis use as a risk factor for schizophrenia in Swedish conscripts of 1969: Historical cohort study. *BMJ* 325 (7374), 1199.

Zammit, S., Owen, M. J., Evans, J., Heron, J., and Lewis, G. 2011. Cannabis, COMT and psychotic experiences. *British Journal of Psychiatry* 199 (5), 380–385. doi: 10.1192/bjp.bp.111.091421

Zammit, S., Spurlock, G., Williams, H., Norton, N., Williams, N., O'Donovan, M. C., et al. 2007. Genotype effects of CHRNA7, CNR1 and COMT in schizophrenia: Interactions with tobacco and cannabis use. *British Journal of Psychiatry* 191, 402–407. doi: 10.1192/bjp.bp.107.036129

Zangen, A., Solinas, M., Ikemoto, S., Goldberg, S. R., and Wise, R. A. 2006. Two brain sites for cannabinoid reward. *Journal of Neuroscience* 26 (18), 4901–4907. doi: 10.1523/JNEUROSCI.3554-05.2006

Zhang, L. R., Morgenstern, H., Greenland, S., Chang, S. C., Lazarus, P., Teare, M. D., et al. 2015. Cannabis smoking and lung cancer risk: Pooled analysis in the International Lung Cancer Consortium. *International Journal of Cancer* 136 (4), 894–903. doi: 10.1002/ijc.29036

Zuardi, A. W., Cosme, R. A., Graeff, F. G., and Guimaraes, F. S. 1993. Effects of ipsapirone and cannabidiol on human experimental anxiety. *Journal of Psychopharmacology* 7 (1 Suppl), 82–88. doi: 10.1177/026988119300700112

Zuardi, A. W., Crippa, J. A., Hallak, J. E., Pinto, J. P., Chagas, M. H., Rodrigues, G. G., et al. 2009. Cannabidiol for the treatment of psychosis in Parkinson's disease. *Journal of Psychopharmacology* 23 (8), 979–983. doi: 10.1177/0269881108096519

Zuardi, A. W., Finkelfarb, E., Bueno, O. F., Musty, R. E., and Karniol, I. G. 1981. Characteristics of the stimulus produced by the mixture of cannabidiol with delta 9-tetrahydrocannabinol. *Archives Internationales de Pharmacodynamie et de Therapie* 249 (1), 137–146.

Zuardi, A. W., Shirakawa, I., Finkelfarb, E., and Karniol, I. G. 1982. Action of cannabidiol on the anxiety and other effects produced by delta 9-THC in normal subjects. *Psychopharmacology* 76 (3), 245–250.

Index